© 2017
Clement Ampadu
drampadu@hotmail.com

ISBN: 978-1-387-01195-7
ID: 21012131
www.lulu.com

All rights reserved. No part of this publication may be produced or transmitted in any form or by any means, electronic or mechanical, including photocopying and recording, or in any information storage and retrieval system, without the prior written permission of the publisher.

Contents

Preface   3

Dedication   4

1 g-Best Proximity Point Theorems for Implicit g-Weak Multiplicative Contractions   5
1.1   Brief Summary   5
1.2   Preliminaries   5
1.3   Main Results   8
1.4   Exercises   12
1.5   References   13

2 r-Best Proximity Point Theorem for Implicit rth-Order Kannan Proximal Multiplicative Contraction   14
2.1   Brief Summary   14
2.2   Preliminaries   14
2.3   Main Results   17
2.4   Exercises   19
2.5   References   20

3   Best Proximity Point Theorems in Multiplicative S-Metric Space   21
3.1   Brief Summary   21
3.2   Preliminaries   21
3.3   Main Results   24
3.4   Exercises   26
3.5   References   28

4   Best Proximity Point Theorems for Implicit Generalized Proximal C-Multiplicative Contraction with Partial Orders   29
4.1   Brief Summary   29
4.2   Preliminaries   29
4.3   Main Results   31
4.4   Exercises   38
4.5   References   41

Preface

The notion of C-class function was introduced by A.H. Ansari [A.H. Ansari, Note on $\varphi - \psi$-contractive type mappings and related fixed point, The 2nd Regional Conference on Mathematics and Applications, PNU, September 2014, 377-380]. On the other hand, the notion of multiplicative C-class function was initiated by C.B. Ampadu and then jointly by Ampadu and Ansari [Clement Ampadu and Arslan Hojat Ansari, FIXED POINT THEOREMS IN COMPLETE MULTIPLICATIVE METRIC SPACES WITH APPLICATION TO MULTIPLICATIVE ANALOGUE OF C-CLASS FUNCTIONS, JP Journal of Fixed Point Theory and Applications, August 2016, Volume 11, Issue 2, Pages 113 - 124].

For a non-self map $T : A \mapsto B$, $T(A) \cap A \neq \emptyset$ is a necessary but not sufficient condition for the existence of a fixed point of T. If $T(A) \cap A = \emptyset$, then the set of fixed points of T is empty. In such a situation, one often attempts to find an element x which is in some sense closest to Tx. If such a point exists, we call it best proximity point.

In this monograph we have defined the non-self multiplicative version of weakly contractive mappings [Ya. I. Alber and S. Guerr-Delabriere, Principle of weakly contractive maps Hilbert spaces, New Results in Operator Theory and its Applications (I.Gohberg and Yu. Lyubich, eds.), Oper. Theory Adv. Appl., vol. 98, Birkhauser, Basel, 1997, pp. 7–22] implicitly via the multiplicative C-class function of Ampadu and Ansari [Clement Ampadu and Arslan Hojat Ansari, FIXED POINT THEOREMS IN COMPLETE MULTIPLICATIVE METRIC SPACES WITH APPLICATION TO MULTIPLICATIVE ANALOGUE OF C-CLASS FUNCTIONS, JP Journal of Fixed Point Theory and Applications, August 2016, Volume 11, Issue 2, Pages 113 - 124], and obtained some sufficient conditions that assure the existence and/or uniqueness of the best proximity point in the multiplicative analogue of Metric space (Chapter 1 and Chapter 2), S-Metric Space (Chapter 3), and Metric space with Partial Order (Chapter 4)

A nice feature of this monograph are the (publishable) exercise set, which begs the reader to explore the beautiful connection between the non-self version of weakly contractive mappings, c-class function, and their multiplicative analogue. The reader will find the notion of multiplicative metric space [A.E. Bashirov, E.M. Kurpnar and A. Ozyapc, Multiplicative calculus and its applications J. Math. Anal. Appl., 337 (2008), 36-48] useful as he or she begins his or her own investigative inquiry.

Prof.Clement Boateng Ampadu

Dedication

Thanking Yahweh, I dedicate this monograph to those who read it, including family, friends, and love ones .

On the occasion of my birth month
Prof.Clement Boateng Ampadu
June, 2017

Chapter 1

g-Best Proximity Point Theorems for Implicit g-Weak Multiplicative Contractions

1.1 Brief Summary

> **Abstract A.1 1**
>
> In this chapter we obtain some best proximity point theorems for so-called g-weak multiplicative contractions defined implicitly using the multiplicative C-class functions of Ampadu and Ansari [Clement Ampadu and Arslan Hojat Ansari, FIXED POINT THEOREMS IN COMPLETE MULTIPLICATIVE METRIC SPACES WITH APPLICATION TO MULTIPLICATIVE ANALOGUE OF C-CLASS FUNCTIONS, JP Journal of Fixed Point Theory and Applications, August 2016, Volume 11, Issue 2, Pages 113 - 124]

1.2 Preliminaries

> **Notation A.1 1**
>
> Let A and B be two nonempty subsets of a multiplicative metric space (X, m). A_0 will denote the set
>
> $$\{x \in A : m(x, y) = m(A, B) \text{ for some } y \in B\}$$
>
> and B_0 will denote the set
>
> $$\{y \in B : m(x, y) = m(A, B) \text{ for some } x \in A\}$$
>
> where $m(A, B) = \inf\{m(x, y) : x \in A \text{ and } y \in B\}$

> **Notation A.2 1**
>
> Ψ will denote the set of all nondecreasing functions $\psi : [1, \infty) \mapsto [1, \infty)$ satisfying the following
>
> (a) $\lim_{n \to \infty} \psi^n(t) = 1$ for all $t > 1$, where ψ^n is the nth iterate of ψ
>
> (b) $\psi(t) < t$, for all $t > 1$; $\psi(1) = 1$, ψ is continuous at $t = 1$

CHAPTER 1. G-BEST PROXIMITY POINT THEOREMS FOR IMPLICIT G-WEAK MULTIPLICATIVE CONTRACTIONS

> **Notation A.3 1**
>
> We denote by Φ the set of all nondecreasing functions $\phi : [1, \infty) \mapsto [1, \infty)$ such that $\phi(t) = 1$ iff $t = 1$. If $\phi \in \Phi$ is continuous at $t = 1$, then we will denote the class of all such functions by Φ_c.

Taking inspiration from [W. Sanhan, C. Mongkolkeha, and P. Kumam, "Generalized proximal ψ-contraction mappings and best proximity points," Abstract and Applied Analysis, vol. 2012, Article ID 896912, 19 pages, 2012] we introduce the following

> **Definition A.4 1**
>
> Let A and B be two nonempty subsets of a multiplicative metric space (X, m). A non-self mapping $T : A \mapsto B$ will be called a proximal ψ-multiplicative contraction of the first kind if $m(u, Tx) = m(A, B) = m(v, Ty)$ implies $m(u, v) \leq \psi(m(x, y))$ for all $u, v, x, y \in A$, where $\psi \in \Psi$. If $\psi(t) = t^\alpha$ for some $\alpha \in [0, 1)$, then we will say T is a proximal multiplicative contraction of the first kind

Taking inspiration from [W. Sanhan, C. Mongkolkeha, and P. Kumam, "Generalized proximal ψ-contraction mappings and best proximity points," Abstract and Applied Analysis, vol. 2012, Article ID 896912, 19 pages, 2012] we introduce the following

> **Definition A.5 1**
>
> Let A and B be two nonempty subsets of a multiplicative metric space (X, m). A non-self mapping $T : A \mapsto B$ will be called a proximal ψ-multiplicative contraction of the second kind if $m(u, Tx) = m(A, B) = m(v, Ty)$ implies $m(Tu, Tv) \leq \psi(m(Tx, Ty))$ for all $u, v, x, y \in A$, where $\psi \in \Psi$. If $\psi(t) = t^\alpha$ for some $\alpha \in [0, 1)$, then we will say T is a proximal multiplicative contraction of the second kind

Taking inspiration from [S. Sadiq Basha, "Best proximity point theorems: an exploration of a common solution to approximation and optimization problems," Applied Mathematics and Computation, vol. 218, no. 19, pp. 9773–9780, 2012] we introduce the following

> **Definition A.6 1**
>
> Let A and B be two nonempty subsets of a multiplicative metric space (X, m). A non-self mapping $T : A \mapsto B$ will be called a weak proximal $F - \phi$-multiplicative contraction of the first kind if $m(u, Tx) = m(A, B) = m(v, Ty)$ implies $m(u, v) \leq F\left[\frac{m(x,y)}{\phi(m(x,y))}\right]$ for all $u, v, x, y \in A$, where $\phi \in \Phi$, and $F(x, y) := F(\frac{x}{y})$ is a multiplicative C-class function [Clement Ampadu and Arslan Hojat Ansari, FIXED POINT THEOREMS IN COMPLETE MULTIPLICATIVE METRIC SPACES WITH APPLICATION TO MULTIPLICATIVE ANALOGUE OF C-CLASS FUNCTIONS, JP Journal of Fixed Point Theory and Applications, August 2016, Volume 11, Issue 2, Pages 113 - 124]

Taking inspiration from [S. Sadiq Basha, "Best proximity point theorems: an exploration of a common solution to approximation and optimization problems," Applied Mathematics and Computation, vol. 218, no. 19, pp. 9773–9780, 2012] we introduce the following

> **Definition A.7 1**
>
> Let A and B be two nonempty subsets of a multiplicative metric space (X, m). A non-self mapping $T : A \mapsto B$ will be called a weak proximal $F - \phi$-multiplicative contraction of the second kind if $m(u, Tx) = m(A, B) = m(v, Ty)$ implies $m(Tu, Tv) \leq F\left[\frac{m(Tx,Ty)}{\phi(m(Tx,Ty))}\right]$ for all $u, v, x, y \in A$, where $\phi \in \Phi$, and $F(x, y) := F(\frac{x}{y})$ is a multiplicative C-class function [Clement Ampadu and Arslan Hojat Ansari, FIXED POINT THEOREMS IN COMPLETE MULTIPLICATIVE METRIC SPACES WITH APPLICATION TO MULTIPLICATIVE ANALOGUE OF C-CLASS FUNCTIONS, JP Journal of Fixed Point Theory and Applications, August 2016, Volume 11, Issue 2, Pages 113 - 124]

> **Definition A.8 1**
>
> Let (X, m) be a multiplicative metric space, and let A, B be two nonempty subsets of X, and let $g : A \mapsto A$. A non-self mapping $T : A \mapsto B$ will be called F-g-weakly multiplicative contractive if there exists $\phi \in \Phi_c$ such that $m(Tx, Ty) \leq F\left[\frac{m(gx,gy)}{\phi(m(gx,gy))}\right]$ for all $x, y \in A$, where $F(x, y) := F(\frac{x}{y})$ is a multiplicative C-class function [Clement Ampadu and Arslan Hojat Ansari, FIXED POINT THEOREMS IN COMPLETE MULTIPLICATIVE METRIC SPACES WITH APPLICATION TO MULTIPLICATIVE ANALOGUE OF C-CLASS FUNCTIONS, JP Journal of Fixed Point Theory and Applications, August 2016, Volume 11, Issue 2, Pages 113- 124]

> **Remark A.9 1**
>
> By the properties of F, we have $m(Tx, Ty) \leq F\left[\frac{m(gx,gy)}{\phi(m(gx,gy))}\right] < m(gx, gy)$ if $x, y \in A$ with $gx \neq gy$, and we say T is a g-multiplicative contractive mapping

Taking inspiration from [V. Sankar Raj, "A best proximity point theorem for weakly contractive non-self-mappings," Nonlinear Analysis, Theory, Methods and Applications, vol. 74, no. 14, pp. 4804–4808, 2011] we introduce the following

> **Definition A.10 1**
>
> Let (A, B) be a pair of nonempty subsets of a multiplicative metric space (X, m) with $A_0 \neq \emptyset$. We will say the pair (A, B) have the P^*-property iff $m(x_1, y_1) = m(A, B) = m(x_2, y_2)$ implies $m(x_1, x_2) = m(y_1, y_2)$, where $x_1, x_2 \in A_0$ and $y_1, y_2 \in B_0$

Taking inspiration from [J. Zhang, Y. Su, and Q. Cheng, "A note on 'A best proximity point theorem for Geraghty-contractions'," Fixed Point Theory and Applications, vol. 2013, p. 99, 2013] we introduce the following

> **Definition A.11 1**
>
> Let (A, B) be a pair of nonempty subsets of a multiplicative metric space (X, m) with $A_0 \neq \emptyset$. We will say the pair (A, B) have the weak P^*-property iff $m(x_1, y_1) = m(A, B) = m(x_2, y_2)$ implies $m(x_1, x_2) \leq m(y_1, y_2)$, where $x_1, x_2 \in A_0$ and $y_1, y_2 \in B_0$

> **Remark A.12 1**
>
> For a nonempty subset A of X, the pair (A, A) has the P^*-property

> **Notation A.13 1**
>
> Let A and B be two nonempty subsets of multiplicative metric space (X, m). Let $g : A \mapsto A$ be a self-mapping and $T : A \mapsto B$ be a non-self mapping
>
> (a) G_A will denote the class of all continuous functions g such that $m(x, y) \leq m(gx, gy)$ for all $x, y \in A$
>
> (b) T_g will denote the class of all functions T such that $m(Tx, Ty) \leq m(Tgx, Tgy)$ for all $x, y \in A$

Taking inspiration from [A. Fernandez-Leon, "Best proximity points for proximal contractions," http://arxiv.org/abs/1207.4349, 2012] we introduce the following

> **Definition A.14 1**
>
> Let A and B be two nonempty subsets of multiplicative metric space (X, m). Let $g : A \mapsto A$ be a self-mapping and $T : A \mapsto B$ be a non-self mapping. We say T preserves multiplicative distance with respect to g if $m(Tgx, Tgy) = m(Tx, Ty)$ for all $x, y \in A$

1.3 Main Results

> **Theorem A.1 1**
>
> Let A and B be two nonempty subsets of a complete multiplicative metric space (X, m). Suppose that A_0 is nonempty and closed. Assume also that the mappings $T : A \mapsto B$ and $g : A \mapsto A$ satisfy the following conditions
>
> (a) T is weak proximal $F - \phi$-multiplicative contraction of the first kind
>
> (b) $g \in G_{A_0}$
>
> (c) $T(A_0) \subseteq B_0$
>
> (d) $A_0 \subseteq g(A_0)$
>
> Then there exists a unique point $x \in A_0$ such that $m(gx, Tx) = m(A, B)$. Moreover, for every $x_0 \in A_0$, there exists a sequence $\{x_n\} \subseteq A$ such that $m(gx_{n+1}, Tx_n) = m(A, B)$ for every $n \in \mathbb{N} \cup \{0\}$ and $x_n \to x$

Proof of Theorem A.1 1

Let $x_0 \in A_0$. Since $T(A_0) \subseteq B_0$ and $A_0 \subseteq g(A_0)$, there exists $x_1 \in A_0$ such that $m(gx_1, Tx_0) = m(A, B)$. Again for $x_1 \in A_0$, there exists $x_2 \in A_0$ such that $m(gx_2, Tx_1) = m(A, B)$. By repeating this process for $x_n \in A_0$ we can find $x_{n+1} \in A_0$ such that $m(gx_{n+1}, Tx_n) = m(A, B)$ for all $n \in \mathbb{N} \cup \{0\}$. Since $T : A \mapsto B$ is a weak proximal $F - \phi$-multiplicative contraction of the first kind and $g \in G_{A_0}$, by the properties of F, we deduce the following

$$m(x_{n+1}, x_n) \leq m(gx_{n+1}, gx_n)$$
$$\leq F\left[\frac{m(x_n, x_{n-1})}{\phi(m(x_n, x_{n-1}))}\right]$$
$$\leq m(x_n, x_{n-1})$$

for every $n \in \mathbb{N}$. Let $t_n = m(x_n, x_{n+1})$, then t_n is bounded non-increasing sequence of nonnegative real numbers. Therefore, t_n converges to t, where $t \geq 1$. Now we claim that $t = 1$. Suppose not, suppose $t > 1$. Since $\phi \in \Phi$, we get $1 < \phi(t) \leq \phi(t_n)$, for all $n \in \mathbb{N}$. Now observe that

$$t_n = m(x_n, x_{n+1})$$
$$\leq m(gx_n, gx_{n+1})$$
$$\leq F\left[\frac{m(x_{n-1}, x_n)}{\phi(m(x_{n-1}, x_n))}\right]$$
$$\leq F\left[\frac{t_{n-1}}{\phi(t_{n-1})}\right]$$
$$\leq F\left[\frac{t_{n-1}}{\phi(t)}\right]$$

Inductively, we obtain $t_{n+p} \leq F\left[\frac{t_n}{\phi(t)^p}\right]$, which is a contradiction for p large enough. Therefore, $t = 1$, and hence $m(x_n, x_{n+1}) \to 1$. Now we claim that $\{x_n\}$ is a multiplicative Cauchy sequence. Suppose not, then there exists $\epsilon > 1$ and subsequences $\{x_{v_k}\}$ and $\{x_{n_k}\}$ of $\{x_n\}$ such that $r_k = m(x_{v_k}, x_{n_k}) \geq \epsilon$, $m(x_{v_k}, x_{n_k-1}) < \epsilon$, and $n_k > v_k \geq k$, for all $k \in \mathbb{N}$. Thus, we have

$$\epsilon \leq r_k$$
$$\leq m(x_{v_k}, x_{n_k-1}) \cdot m(x_{n_k-1}, x_{n_k})$$
$$< \epsilon \cdot t_{n_k-1}$$

Now taking limits in the above, we deduce $\lim_{k \to \infty} r_k = \epsilon$. Since $m(gx_{v_k+1}, Tx_{v_k}) = m(A, B)$ and $m(gx_{n_k+1}, Tx_{n_k}) = m(A, B)$, and T is a weak proximal $F - \phi$-multiplicative contraction of the first kind, we deduce that

$$m(x_{v_k+1}, x_{n_k+1}) \leq m(gx_{v_k+1}, gx_{n_k+1})$$
$$\leq F\left[\frac{m(x_{v_k}, x_{n_k})}{\phi(m(x_{v_k}, x_{n_k}))}\right]$$

Proof of Theorem A.1 continued 1

It follows that

$$\begin{aligned}
\epsilon &\leq r_k \\
&\leq m(x_{v_k}, x_{v_k+1}) \cdot m(x_{v_k+1}, x_{n_k+1}) \cdot m(x_{n_k+1}, x_{n_k}) \\
&= t_{v_k} \cdot t_{n_k} \cdot m(x_{v_k+1}, x_{n_k+1}) \\
&\leq t_{v_k} \cdot t_{n_k} \cdot F\left[\frac{m(x_{v_k}, x_{n_k})}{\phi(m(x_{v_k}, x_{n_k}))}\right] \\
&\leq t_{v_k} \cdot t_{n_k} \cdot F\left[\frac{m(x_{v_k}, x_{n_k})}{\phi(\epsilon)}\right]
\end{aligned}$$

Now taking limits in the above as $k \to \infty$, we deduce that $\epsilon \leq F\left[\frac{\epsilon}{\phi(\epsilon)}\right]$, which is a contradiction. Therefore $\{x_n\}$ is a multiplicative Cauchy sequence. By the completeness of X and since A_0 is closed, we have $x_n \to x \in A_0$. Moreover, by continuity of g, we have $gx_n \to gx$ and thus $gx \in A_0$, since $gx_n \in A_0$ for all $n \in \mathbb{N}$. On the other hand since $x \in A_0$ and $T(A_0) \subseteq B_0$, there exists $z \in A_0$ such that $m(z, Tx) = m(A, B)$. Now observe by the properties of F and since T is a weak proximal $F - \phi$-multiplicative contraction of the first kind, we deduce that

$$\begin{aligned}
m(z, gx_{n+1}) &\leq F\left[\frac{m(x, x_n)}{\phi(m(x, x_n))}\right] \\
&\leq m(x, x_n)
\end{aligned}$$

Now taking limits in the above, we get $\lim_{n \to \infty} m(z, gx_{n+1}) = 1$, and thus, $z = gx$. It follows that $m(gx, Tx) = m(A, B)$. As for uniqueness suppose that $x^* \in A_0$ is such that $m(gx^*, Tx^*) = m(A, B)$ and $x \neq x^*$. Since $g \in G_{A_0}$, and T is a weak proximal $F - \phi$-multiplicative contraction of the first kind, we deduce that

$$\begin{aligned}
m(x, x^*) &\leq m(gx, gx^*) \\
&\leq F\left[\frac{m(x, x^*)}{\phi(m(x, x^*))}\right] \\
&\leq m(x, x^*)
\end{aligned}$$

which is a contradiction, thus $x = x^*$

Remark A.2 1

The above theorem still holds if g is the identity

Theorem A.3 1

Let A and B be closed subsets of a complete multiplicative metric space (X, m) such that $A_0, B_0 \neq \emptyset$ and the pair (A, B) has the weak P^*-property. Suppose that the mappings $g : A \mapsto A$ and $T : A \mapsto B$ satisfy the following conditions

(a) T is a F-g weak multiplicative contraction

(b) $T(A_0) \subset B_0$

(c) $A_0 \subset g(A_0)$

Then there exists an element $x^* \in A_0$ such that $m(gx^*, Tx^*) = m(A, B)$. Further if g is 1-1, we have a unique element $x^* \in A$ such that $m(gx^*, Tx^*) = m(A, B)$

Proof of Theorem A.3 1

Let x_0 be an element of A_0. Since $T(A_0) \subset B_0$ and $A_0 \subset g(A_0)$ it follows that there exists $x_1 \in A_0$ such that $m(gx_1, Tx_0) = m(A, B)$. Again since $T(A_0) \subset B_0$ and $A_0 \subset g(A_0)$ it follows that there exists $x_2 \in A_0$ such that $m(gx_2, Tx_1) = m(A, B)$. Continuing this process, we can find a sequence $\{x_n\}$ in A_0 such that $m(gx_n, Tx_{n-1}) = m(A, B)$ for all $n \in \mathbb{N}$. Since (A, B) has the weak P^*-property we conclude that $m(gx_n, gx_{n+1}) \leq m(Tx_{n-1}, Tx_n)$ for all $n \in \mathbb{N}$. Now as T is a F-g-weak multiplicative contraction, we deduce that

$$m(gx_n, gx_{n+1}) \leq m(Tx_{n-1}, Tx_n)$$
$$\leq F\left[\frac{m(gx_{n-1}, gx_n)}{\phi(m(gx_{n-1}, gx_n))}\right]$$

where $\phi \in \Phi_c$. Now set $t_n = m(gx_n, gx_{n+1})$, then $\{t_n\}$ is non-increasing sequence of nonnegative real numbers and hence multiplicative converges. Let $t \geq 1$ be the limit of the sequence $\{t_n\}$. We claim that $t = 1$. If not, since ϕ is nondecreasing function, we deduce that $\phi(t_n) \geq \phi(t) > 1$, for all $n \in \mathbb{N}$. Thus for any positive integer n, we deduce that $t_{n+1} \leq F\left[\frac{t_n}{\phi(t)}\right]$. Now for all $t_1 < \phi(t)^n$, we obtain that $t_{n+1} \leq F\left[\frac{t_1}{\phi(t)^n}\right] < 1$, which is a contradiction, therefore $t = 1$ and the sequence $\{m(gx_n, gx_{n+1})\}$ converges to one. Since

$$m(gx_n, gx_{n+1}) \leq m(Tx_{n-1}, Tx_n) \leq m(gx_{n-1}, gx_n)$$

it follows that the sequence $\{m(Tx_{n-1}, Tx_n)\}$ converges to one. Now we show that $\{Tx_n\}$ is a multiplicative Cauchy sequence. Let $\epsilon > 1$ be given and choose $n(\epsilon)$ such that $m(Tx_n, Tx_{n+1}) \leq \min\{\sqrt{\epsilon}, \phi(\sqrt{\epsilon})\}$ for all $n \geq n(\epsilon)$. Fix $n \geq n(\epsilon)$ and let

$$A(n, \epsilon) = \{x \in A : m(Tx_n, Tx) \leq \epsilon\}$$

If $x \in A(n, \epsilon)$ and $u \in A$ is such that $m(gu, Tx) = m(A, B)$, then $u \in A(n, \epsilon)$. Now since $m(gx_{n+1}, Tx_n) = m(A, B)$, then by the weak P^*-property, we have $m(gx_{n+1}, gu) \leq m(Tx_n, Tx)$, and two cases must be considered to establish this fact.

Case 1: $m(gx_{n+1}, gu) \leq \sqrt{\epsilon}$

We observe by the properties of F that

$$m(Tx_n, Tu) \leq m(Tx_n, Tx_{n+1}) \cdot m(Tx_{n+1}, Tu)$$
$$\leq \sqrt{\epsilon} \cdot F\left[\frac{m(gx_{n+1}, gu)}{\phi(m(gx_{n+1}, gu))}\right]$$
$$\leq \sqrt{\epsilon} \cdot m(gx_{n+1}, gu)$$
$$\leq \epsilon$$

Case 2: $\sqrt{\epsilon} \leq m(gx_{n+1}, gu) < \epsilon$

We observe by the properties of F that

$$m(Tx_n, Tu) \leq m(Tx_n, Tx_{n+1}) \cdot m(Tx_{n+1}, Tu)$$
$$\leq \phi(\sqrt{\epsilon}) \cdot F\left[\frac{m(gx_{n+1}, gu)}{\phi(m(gx_{n+1}, gu))}\right]$$
$$\leq \phi(\sqrt{\epsilon}) \cdot F\left[\frac{m(gx_{n+1}, gu)}{\phi(\sqrt{\epsilon})}\right]$$
$$\leq \phi(\sqrt{\epsilon}) \cdot m(gx_{n+1}, gu)$$
$$\leq m(gx_{n+1}, gu)$$
$$\leq \epsilon$$

> **Proof of Theorem A.3 Continued 1**
>
> It follows that $u \in A(n,\epsilon)$. Now we prove that $x_{n+k} \in A(n,\epsilon)$ for all $k \geq 1$. From $x_n \in A(n,\epsilon)$ and $m(gx_{n+1}, Tx_n) = m(A,B)$, we deduce that $x_{n+1} \in A(n,\epsilon)$. Now assume that $x_{n+k} \in A(n,\epsilon)$ holds for some $k \geq 1$. From $x_{n+k} \in A(n,\epsilon)$ and $m(gx_{n+k+1}, Tx_{n+k}) = m(A,B)$, we deduce that $x_{n+k+1} \in A(n,\epsilon)$, and it follows that $x_{n+k} \in A(n,\epsilon)$ holds for all $k \geq 1$. Hence $\{Tx_n\}$ is a multiplicative Cauchy sequence. From the multiplicative completeness of the space X, the sequence $\{Tx_n\}$ converges to some element $y^* \in B$. From $m(gx_{n+1}, gx_{m+1}) \leq m(Tx_n, Tx_m)$, we deduce that the sequence $\{gx_n\}$ is also a multiplicative Cauchy sequence. As A is a multiplicative subspace of X, then there exists $z \in A$ such that $gx_n \to z$. Therefore we have $m(z, y^*) = \lim_{n \to \infty} m(gx_{n+1}, Tx_n) = m(A,B)$, and so $z \in A_0$. Since A_0 is contained in $g(A_0)$, there exists $x^* \in A_0$ such that $z = gx^*$. Since $T(A_0) \subset B_0$, there is an element $\Xi \in A_0$ such that $m(g\Xi, Tx^*) = m(A,B)$. Since T is an F-g weak multiplicative contraction, (A,B) has the weak P^*-property, and ϕ is continuous at $t=1$, we deduce that
>
> $$m(gx_{n+1}, g\Xi) \leq m(Tx_n, Tx^*)$$
> $$\leq F\left[\frac{m(gx_{n+1}, gx^*)}{\phi(m(gx_{n+1}, gx^*))}\right]$$
>
> Taking limits in the above as $n \to \infty$, we deduce that $g\Xi = gx^*$. It follows that $m(gx^*, Tx^*) = m(A,B)$. As for uniqueness suppose that $z^* \in A$ is such that $m(gz^*, Tz^*) = m(A,B)$, then we deduce that
>
> $$m(gx^*, gz^*) \leq m(Tx^*, Tz^*)$$
> $$\leq F\left[\frac{m(gx^*, gz^*)}{\phi(m(gx^*, gz^*))}\right]$$
>
> From the above we deduce that $gx^* = gz^*$ and hence $z^* \in g^{-1}gx^*$. If g is 1-1, uniqueness follows

> **Remark A.4 1**
>
> If g is the identity in the previous theorem, then it still holds. Moreover, we get a multiplicative generalization of a result due to Rhoades [B. E. Rhoades, "Some theorems on weakly contractive maps," Nonlinear Analysis, Theory, Methods and Applications, vol. 47, no. 4, pp. 2683–2693, 2001]

1.4 Exercises

> **Exercise A.1 1**
>
> Let A and B be two nonempty subsets of a complete multiplicative metric space (X, m). Suppose that $T(A_0)$ is nonempty and closed. Assume also that the mappings $T : A \mapsto B$ and $g : A \mapsto A$ satisfy the following conditions
>
> (a) T is weak proximal $F - \phi$-multiplicative contraction of the second kind
>
> (b) $T \in T_G$
>
> (c) $T(A_0) \subseteq B_0$
>
> (d) $A_0 \subseteq g(A_0)$
>
> Then there exists a unique point $x \in A_0$ such that $m(gx, Tx) = m(A,B)$. Moreover, if T is injective on A, then the point x such that $m(gx, Tx) = m(A,B)$ is unique.

> **Exercise A.2 1**
>
> Taking inspiration from [Maryam A. Alghamdi, Naseer Shahzad, and Francesca Vetro, Best Proximity Points for Some Classes of Proximal Contractions, Abstract and Applied Analysis Volume 2013, Article ID 713252, 10 pages] give an example showing that the weak P^*-property in Theorem A.3 cannot be relaxed

1.5 References

(1) Clement Ampadu and Arslan Hojat Ansari, FIXED POINT THEOREMS IN COMPLETE MULTIPLICATIVE METRIC SPACES WITH APPLICATION TO MULTIPLICATIVE ANALOGUE OF C-CLASS FUNCTIONS, JP Journal of Fixed Point Theory and Applications, August 2016, Volume 11, Issue 2, Pages 113 - 124

(2) W. Sanhan, C. Mongkolkeha, and P. Kumam, "Generalized proximal ψ-contraction mappings and best proximity points," Abstract and Applied Analysis, vol. 2012, Article ID 896912, 19 pages, 2012

(3) S. Sadiq Basha, "Best proximity point theorems: an exploration of a common solution to approximation and optimization problems," Applied Mathematics and Computation, vol. 218, no. 19, pp. 9773–9780, 2012

(4) V. Sankar Raj, "A best proximity point theorem for weakly contractive non-self-mappings," Nonlinear Analysis, Theory, Methods and Applications, vol. 74, no. 14, pp. 4804–4808, 2011

(5) J. Zhang, Y. Su, and Q. Cheng, "A note on 'A best proximity point theorem for Geraghty-contractions'," Fixed Point Theory and Applications, vol. 2013, p. 99, 2013

(6) A. Fernandez-Leon, "Best proximity points for proximal contractions," http://arxiv.org/abs/1207.4349, 2012

(7) B. E. Rhoades, "Some theorems on weakly contractive maps," Nonlinear Analysis, Theory, Methods and Applications, vol. 47, no. 4, pp. 2683–2693, 2001

(8) Maryam A. Alghamdi, Naseer Shahzad, and Francesca Vetro, Best Proximity Points for Some Classes of Proximal Contractions, Abstract and Applied Analysis Volume 2013, Article ID 713252, 10 pages

Chapter 2

r-Best Proximity Point Theorem for Implicit rth-Order Kannan Proximal Multiplicative Contraction

2.1 Brief Summary

Abstract B.1 1

Motivated by the concepts of non-self weakly contractive mappings [V. Sankar Raj, A best proximity theorem for weakly contractive non-self mappings, Nonlinear Anal. 74 (2011), 4804-4808], proximal contraction [S. Sadiq Basha, Best Proximity Points: Optimal Solutions, J. Optim. Theory Appl. 151 (2011) no.1, 210-216], generalized proximal contraction [S. Sadiq Basha,Best proximity point theorems: resolution of an important non-linear programming problem, Optim. Lett. 7 (2013), no.6, 1167-1177], and higher-order Kannan contraction[Clement Boateng Ampadu. Higher-Order Kannan Mapping Theorem in Metric Spaces. Unpublished]. We introduce a concept of implicit higher-order Kannan proximal multiplicative contraction and prove r-best proximity point theorem for this multiplicative contraction.

2.2 Preliminaries

Taking inspiration from [V. Sankar Raj, A best proximity theorem for weakly contractive non-self mappings, Nonlinear Anal. 74 (2011), 4804-4808] we introduce the following

Definition B.1 1

Let A and B be nonempty subsets of a multiplicative metric space (X, m). A mapping $T : A \mapsto B$ will be called a higher-order implicit weak multiplicative contraction if it satisfies $m(T^r x, T^r y) \leq F\left[\frac{m(x,y)}{\phi(m(x,y))}\right]$ for all $x, y \in X$ and $r \in \mathbb{N}$, where $\phi : [1, \infty) \mapsto [1, \infty)$ is a continuous and nondecreasing function such that ϕ is positive on $(1, \infty)$, $\phi(1) = 1$, and $\lim_{t \to \infty} \phi(t) = \infty$, and $F(x, y) := F\left(\frac{x}{y}\right)$ is a multiplicative C-class function [Clement Ampadu and Arslan Hojat Ansari, FIXED POINT THEOREMS IN COMPLETE MULTIPLICATIVE METRIC SPACES WITH APPLICATION TO MULTIPLICATIVE ANALOGUE OF C-CLASS FUNCTIONS, JP Journal of Fixed Point Theory and Applications, August 2016, Volume 11, Issue 2, Pages 113 - 124]

Definition B.2 1

An element $x \in A$ will be called an r-fixed point of the map $T: A \mapsto B$ if $T^r x = x$ for any $r \in \mathbb{N}$

Definition B.3 1

Let (X, m) be a multiplicative metric space. We say $x \in A$ is an r-best proximity point of T if $m(x, T^r x) = multdist(A, B)$ for any $r \in \mathbb{N}$, where $multdist(A, B) = \inf\{m(A, B) : x \in A, y \in B\}$

Notation B.4 1

A_0 will denote the set $\{x \in A : m(x, y) = multdist(A, B) \text{ for some } y \in B\}$

Notation B.5 1

B_0 will denote the set $\{y \in A : m(x, y) = multdist(A, B) \text{ for some } x \in A\}$

Taking inspiration from [V. Sankar Raj, A best proximity theorem for weakly contractive non-self mappings, Nonlinear Anal. 74 (2011), 4804-4808] we introduce the following

Definition B.6 1

Let A and B be two nonempty subsets of a multiplicative metric space (X, m) with $A_0 \neq \emptyset$. We will say the pair (A, B) have the P^*-property iff $m(x_1, y_1) = multdist(A, B)$ and $m(x_2, y_2) = multdist(A, B)$ implies $m(x_1, x_2) = m(y_1, y_2)$, where $x_1, x_2 \in A_0$ and $y_1, y_2 \in B_0$

Remark B.7 1

For any nonempty subset A of X, the pair (A, A) has the P^*-property

Proposition B.8 1

[Jeffery Ezearn, Higher-order Lipschitz Mappings, Fixed Point Theory and Applications (2015) 2015:88] Let (X, d) be a metric space, and let T be an rth-order Banach contraction mapping on X. For every pair $x \neq y \in X$, define

$$Z := Z(x, y) = \max_{0 \leq v \leq r-1} \beta^{-v} \frac{d(T^v x, T^v y)}{d(x, y)}$$

then

$$Z = \max_{n \in 0 \cup \mathbb{N}} \beta^{-n} \frac{d(T^n x, T^n y)}{d(x, y)}$$

where $\beta \in [0, 1)$

Taking inspiration from [S. Sadiq Basha, Best Proximity Points: Optimal Solutions, J. Optim. Theory Appl. 151 (2011) no.1, 210-216] we introduce the following

Definition B.9 1

A map $T: A \mapsto B$ will be called a higher-order proximal multiplicative contraction if there exists $Z \geq 1$ and $\beta \in [0, 1)$ given by the previous proposition such that for all $u_1, u_2, x_1, x_2 \in A$, $m(u_1, T^r x_1) = multdist(A, B)$ and $m(u_2, T^r x_2) = multdist(A, B)$ implies $m(u_1, u_2) \leq m(x_1, x_2)^{Z\beta^r}$ for any $r \in \mathbb{N}$

> **Remark B.10 1**
>
> If $A = B$ in the previous definition, then we get the multiplicative version of the higher-order Banach contraction in the sense of [Clement Ampadu(2015). Generalization of Higher Order Contraction Mapping Theorem. Unpublished]

Taking inspiration from [S. Sadiq Basha and N. Shahzad, Best proximity point theorems for generalized proximal contractions, Fixed Point Theory Appl., 2012(2012), Article ID 42] we introduce the following

> **Definition B.11 1**
>
> Let (X, m) be a multiplicative metric space. The set B will be called approximately compact with respect to A if every sequence $\{y_n\}$ of B satisfying the condition that $m(x, y_n)$ multiplicative converges to $m(x, B)$ for some $x \in A$ has a multiplicative convergent subsequence

> **Remark B.12 1**
>
> Any compact set is approximately compact, and any set is approximately compact with respect to itself. Further if A is compact and B is approximately compact with respect to A, then A_0 and B_0 are nonempty

Taking inspiration from [S. Sadiq Basha, Best proximity point theorems: resolution of an important non-linear programming problem, Optim. Lett. 7 (2013), no.6, 1167-1177] we introduce the following

> **Definition B.13 1**
>
> A mapping $T : A \mapsto B$ will be called an implicit higher-order generalized proximal multiplicative contraction, if for all $u_1, u_2, x_1, x_2 \in A$ and $r \in \mathbb{N}$, $m(u_1, T^r x_1) = multdist(A, B)$ and $m(u_2, T^r x_2) = multdist(A, B)$ implies $m(u_1, u_2) \leq F\left[\frac{m(x_1, x_2)}{\phi(m(x_1, x_2))}\right]$, where $\phi : [1, \infty) \mapsto [1, \infty)$ is continuous and nondecreasing and is such that ϕ is positive on $[1, \infty)$, $\phi(1) = 1$, $\lim_{t \to \infty} \phi(t) = \infty$, and $F(x, y) := F\left(\frac{x}{y}\right)$ is a multiplicative C-class function [Clement Ampadu and Arslan Hojat Ansari, FIXED POINT THEOREMS IN COMPLETE MULTIPLICATIVE METRIC SPACES WITH APPLICATION TO MULTIPLICATIVE ANALOGUE OF C-CLASS FUNCTIONS, JP Journal of Fixed Point Theory and Applications, August 2016, Volume 11, Issue 2, Pages 113 - 124]

> **Definition B.14 1**
>
> [Clement Boateng Ampadu. Higher-Order Kannan Mapping Theorem in Metric Spaces. Unpublished] Let (X, d) be a metric space. We say $T : X \mapsto X$ is a higher-order Kannan mapping if $d(T^r x, T^r y) \leq \sum_{q=0}^{r-1} c_q [d(T^q x, T^{q+1} x) + d(T^q y, T^{q+1} y)]$ for all $x, y \in X$, where $0 \leq c_q < \frac{1}{2}$ for all $0 \leq q \leq r - 1$

Proposition B.15 1

[Clement Boateng Ampadu. Higher-Order Kannan Mapping Theorem in Metric Spaces. Unpublished] Let (X,d) be a metric space, and $T : X \mapsto X$ be a higher-order Kannan contraction. For every pair $x \neq y$, define

$$Z := Z(x,y) = \max_{0 \leq t \leq r-1} \beta^{-t} \frac{d(T^t x, T^t y)}{d(x,Tx) + d(y,Ty)}$$

then

$$Z = \max_{n \in \mathbb{N} \cup \{0\}} \beta^{-n} \frac{d(T^n x, T^n y)}{d(x,Tx) + d(y,Ty)}$$

where $\beta \in [0, \frac{1}{2})$

Definition B.16 1

[Clement Boateng Ampadu. Higher-Order Kannan Mapping Theorem in Metric Spaces. Unpublished] Let (X,d) be a metric space. A map $T : X \mapsto X$ will be called a higher-order Kannan contraction if $d(T^r x, T^r y) \leq Z \beta^r [d(x,Tx) + d(y,Ty)]$ for all $x, y \in X$ and all $r \in \mathbb{N}$, where $\beta \in [0, \frac{1}{2})$ and $Z \geq 2$ is the bound given by the previous Proposition

Definition B.17 1

Let A and B be two nonempty subsets of a multiplicative metric space (X,m), and let Z and β be given by the previous proposition. We say $T : A \mapsto B$ is an implicit higher-order Kannan proximal multiplicative contraction if for all $u_1, u_2, x_1, x_2 \in A$ and $r \in \mathbb{N}$, $m(u_1, T^r x_1) = multdist(A,B)$ and $m(u_2, T^r x_2) = multdist(A,B)$ implies $m(u_1, u_2) \leq F\left[\frac{(m(x_1,u_1) \cdot m(x_2,u_2))^{Z\beta^r}}{\phi(m(x_1,u_1), m(x_2,u_2))}\right]$, where $\phi : [1,\infty)^2 \mapsto [1,\infty)$ is a continuous and nondecreasing function such that $\phi(x_1, x_2) = 1$ iff $x_1 = x_2 = 1$, and $F(x,y) := F\left(\frac{x}{y}\right)$ is a multiplicative C-class function [Clement Ampadu and Arslan Hojat Ansari, FIXED POINT THEOREMS IN COMPLETE MULTIPLICATIVE METRIC SPACES WITH APPLICATION TO MULTIPLICATIVE ANALOGUE OF C-CLASS FUNCTIONS, JP Journal of Fixed Point Theory and Applications, August 2016, Volume 11, Issue 2, Pages 113 - 124]

2.3 Main Results

Theorem B.1 1

Let (X,m) be a complete multiplicative metric space, A and B be two nonempty, closed subsets of X such that B is approximately compact with respect to A. Suppose that A_0 and B_0 are nonempty and $T : A \mapsto B$ is a non-self-mapping satisfying the following conditions

(a) T is an implicit higher-order Kannan proximal multiplicative contraction

(b) $T^r(A_0) \subseteq B_0$ for any $r \in \mathbb{N}$

Then there exists an element $x \in A$ such that $m(x, T^r x) = multdist(A,B)$ for any $r \in \mathbb{N}$. Further the sequence $\{x_n\}$ converges to the r-best proximity point x, where for a fixed $x_0 \in A_0$ the sequence $\{x_n\}$ is given by $m(x_{n+1}, T^r x_n) = multdist(A,B)$ for all $n \geq 0$ and $r \in \mathbb{N}$

Proof of Theorem B.1 1

Let x_0 be a fixed element in A_0. Since $T^r(A_0) \subseteq B_0$, $T^r x_0$ is an element of B_0. So by the definition of B_0, there exists an element $x_1 \in A_0$ such that $m(x_1, T^r x_0) = multdist(A, B)$. Again, since $T^r(A_0) \subseteq B_0$, we have $T^r x_1 \in B_0$, it follows there exists $x_2 \in A_0$ such that $m(x_2, T^r x_1) = multdist(A, B)$. Continuing this process, we derive a sequence $\{x_n\}$ in A_0 such that $m(x_{n+1}, T^r x_n) = multdist(A, B)$ for every $n \geq 0$. Since T is an implicit higher-order Kannan proximal contraction, we have,

$$m(x_n, x_{n+1}) \leq F\left[\frac{(m(x_{n-1}, x_n) \cdot m(x_n, x_{n+1}))^{Z\beta^r}}{\phi(m(x_{n-1}, x_n), m(x_n, x_{n+1}))}\right]$$
$$\leq (m(x_{n-1}, x_n) \cdot m(x_n, x_{n+1}))^{Z\beta^r}$$

Consequently, we get $m(x_n, x_{n+1}) \leq m(x_{n-1}, x_n)$, it follows that the sequence $\{m(x_n, x_{n+1})\}$ is a monotone decreasing sequence of real numbers, and so there exists $\mu \geq 1$ such that $\lim_{n \to \infty} m(x_n, x_{n+1}) = \mu$. We claim $\mu = 1$. If not observe by taking limits as $n \to \infty$ in the inequality below

$$m(x_n, x_{n+1}) \leq F\left[\frac{(m(x_{n-1}, x_n) \cdot m(x_n, x_{n+1}))^{Z\beta^r}}{\phi(m(x_{n-1}, x_n), m(x_n, x_{n+1}))}\right]$$

we get $\mu \leq F\left[\frac{(\mu \cdot \mu)^{Z\beta^r}}{\phi(\mu, \mu)}\right]$, and since $Z\beta^r < \frac{1}{2}$ for any $r \in \mathbb{N}$, we get a contradiction unless $\mu = 1$, it follows that $\lim_{n \to \infty} m(x_n, x_{n+1}) = 1$. Now we show that $\{x_n\}$ is a multiplicative Cauchy sequence. Suppose not, then there exists $\epsilon > 1$ such that $n_k > v_k \geq k$ with $m(x_{v_k}, x_{n_k}) \geq \epsilon$ and $m(x_{v_k}, x_{n_k - 1}) < \epsilon$ for each $k \in \mathbb{N}$. Now observe that

$$m(x_{v_k}, x_{n_k}) \leq m(x_{v_k}, x_{n_k - 1}) \cdot m(x_{n_k - 1}, x_{n_k})$$
$$\leq \epsilon \cdot m(x_{n_k - 1}, x_{n_k})$$

Since $\lim_{n \to \infty} m(x_n, x_{n+1}) = 1$, then taking limits in the above as $k \to \infty$, we deduce that $\lim_{k \to \infty} m(x_{v_k}, x_{n_k}) = \epsilon$. Since $m(x_{n+1}, T^r x_n) = multdist(A, B)$, we deduce that $\lim_{k \to \infty} m(x_{n_k + 1}, T^r x_{n_k}) = multdist(A, B)$ and $\lim_{k \to \infty} m(x_{v_k + 1}, T^r x_{v_k}) = multdist(A, B)$. Now observe that since T is an implicit higher-order Kannan proximal multiplicative contraction we have

$$m(x_{n_k + 1}, x_{v_k + 1}) \leq F\left[\frac{(m(x_{n_k}, x_{n_k + 1}) \cdot m(x_{v_k}, x_{v_k + 1}))^{Z\beta^r}}{\phi(m(x_{n_k}, x_{n_k + 1}), m(x_{v_k}, x_{v_k + 1}))}\right]$$

Combining the above inequality, with the following

$$m(x_{n_k}, x_{v_k}) \leq m(x_{n_k}, x_{n_k + 1}) \cdot m(x_{n_k + 1}, x_{v_k + 1}) \cdot m(x_{v_k + 1}, x_{v_k})$$

we deduce that

$$m(x_{n_k}, x_{v_k}) \leq m(x_{n_k}, x_{n_k + 1}) \cdot m(x_{v_k + 1}, x_{v_k}) \cdot F\left[\frac{(m(x_{n_k}, x_{n_k + 1}) \cdot m(x_{v_k}, x_{v_k + 1}))^{Z\beta^r}}{\phi(m(x_{n_k}, x_{n_k + 1}), m(x_{v_k}, x_{v_k + 1}))}\right]$$

Now taking limits in the above as $k \to \infty$ we deduce that $\epsilon \leq F\left[\frac{1}{\phi(1,1)}\right] = 1$, which is a contradiction. Hence it follows that $\{x_n\}$ is a multiplicative Cauchy sequence. Since A is a closed subset of the complete space X, there exists $x \in A$ such that $\lim_{n \to \infty} x_n = x$. Now taking limits in $m(x_{n+1}, T^r x_n) = multdist(A, B)$, and using the continuity of T^r for any $r \in \mathbb{N}$, we deduce that $m(x, T^r x) = multdist(A, B)$

2.4 Exercises

Exercise B.1 1

Prove the following: Let (X, m) be a complete multiplicative metric space, A and B be two nonempty, closed subsets of X such that B is approximately compact with respect to A. Suppose that A_0 and B_0 are nonempty and $T : A \mapsto B$ is a non-self-mapping satisfying the following conditions

(a) T is an implicit higher-order generalized proximal multiplicative contraction

(b) $T^r(A_0) \subseteq B_0$ for any $r \in \mathbb{N}$

Then there exists an element $x \in A$ such that $m(x, T^r x) = multdist(A, B)$ for any $r \in \mathbb{N}$. Further the sequence $\{x_n\}$ converges to the r-best proximity point x, where for a fixed $x_0 \in A_0$ the sequence $\{x_n\}$ is given by $m(x_{n+1}, T^r x_n) = multdist(A, B)$ for all $n \geq 0$ and $r \in \mathbb{N}$

Exercise B.2 1

Taking inspiration from [AJAY SHARMA, BALWANT SINGH THAKUR, BEST PROXIMITY POINTS FOR K-PROXIMAL CONTRACTION, International Journal of Analysis and Applications, Volume 6, Number 1 (2014), 82-88] given an example in support of Theorem B.1

Exercise B.3 1

Let A and B be two nonempty subsets of a multiplicative metric space (X, m), and let J and Γ be given by Exercise A.3 [Clement Ampadu, Characterization Theorems Inspired by the Hardy-Rogers Map I: Some Results in Metric Spaces. lulu.com, 2016. ISBN: 1365101185, 9781365101182]. We say $T : A \mapsto B$ is an implicit higher-order Reich proximal multiplicative contraction if for all $u_1, u_2, x_1, x_2 \in A$ and $r \in \mathbb{N}$, $m(u_1, T^r x_1) = multdist(A, B)$ and $m(u_2, T^r x_2) = multdist(A, B)$ implies $m(u_1, u_2) \leq F\left[\frac{(m(x_1,x_2) \cdot m(x_1,u_1) \cdot m(x_2,u_2))^{J\Gamma^r}}{\phi(m(x_1,x_2), m(x_1,u_1), m(x_2,u_2))}\right]$, where $\phi : [1, \infty)^3 \mapsto [1, \infty)$ is a continuous and nondecreasing function such that $\phi(x_1, x_2, x_3) = 1$ iff $x_1 = x_2 = x_3 = 1$, and $F(x, y) := F(\frac{x}{y})$ is a multiplicative C-class function [Clement Ampadu and Arslan Hojat Ansari, FIXED POINT THEOREMS IN COMPLETE MULTIPLICATIVE METRIC SPACES WITH APPLICATION TO MULTIPLICATIVE ANALOGUE OF C-CLASS FUNCTIONS, JP Journal of Fixed Point Theory and Applications, August 2016, Volume 11, Issue 2, Pages 113 - 124]

Using the above definition, prove the following: Let (X, m) be a complete multiplicative metric space, A and B be two nonempty, closed subsets of X such that B is approximately compact with respect to A. Suppose that A_0 and B_0 are nonempty and $T : A \mapsto B$ is a non-self-mapping satisfying the following conditions

(a) T is an implicit higher-order Reich proximal multiplicative contraction

(b) $T^r(A_0) \subseteq B_0$ for any $r \in \mathbb{N}$

Then there exists an element $x \in A$ such that $m(x, T^r x) = multdist(A, B)$ for any $r \in \mathbb{N}$. Further the sequence $\{x_n\}$ converges to the r-best proximity point x, where for a fixed $x_0 \in A_0$ the sequence $\{x_n\}$ is given by $m(x_{n+1}, T^r x_n) = multdist(A, B)$ for all $n \geq 0$ and $r \in \mathbb{N}$

> **Exercise B.4 1**
>
> Taking inspiration from [S.K. Chatterjee, Fixed point theorems, Comptes. Rend. Acad. Bulgaria Sci. 25(1972), 727-730], Definition B.14, Proposition B.15, Definition B.16, and Definition B.17, introduce a concept of implicit higher-order Chatterjee proximal multiplicative contraction and use it to prove the following: Let (X, m) be a complete multiplicative metric space, A and B be two nonempty, closed subsets of X such that B is approximately compact with respect to A. Suppose that A_0 and B_0 are nonempty and $T : A \mapsto B$ is a non-self-mapping satisfying the following conditions
>
> (a) T is an implicit higher-order Chatterjee proximal multiplicative contraction
>
> (b) $T^r(A_0) \subseteq B_0$ for any $r \in \mathbb{N}$
>
> Then there exists an element $x \in A$ such that $m(x, T^r x) = multdist(A, B)$ for any $r \in \mathbb{N}$. Further the sequence $\{x_n\}$ converges to the r-best proximity point x, where for a fixed $x_0 \in A_0$ the sequence $\{x_n\}$ is given by $m(x_{n+1}, T^r x_n) = multdist(A, B)$ for all $n \geq 0$ and $r \in \mathbb{N}$

2.5 References

(1) V. Sankar Raj, A best proximity theorem for weakly contractive non-self mappings, Nonlinear Anal. 74 (2011), 4804-4808

(2) S. Sadiq Basha, Best Proximity Points: Optimal Solutions, J. Optim. Theory Appl. 151 (2011) no.1, 210-216

(3) S. Sadiq Basha, Best proximity point theorems: resolution of an important non-linear programming problem, Optim. Lett. 7 (2013), no.6, 1167-1177

(4) Clement Boateng Ampadu. Higher-Order Kannan Mapping Theorem in Metric Spaces. Unpublished

(5) Clement Ampadu and Arslan Hojat Ansari, FIXED POINT THEOREMS IN COMPLETE MULTIPLICATIVE METRIC SPACES WITH APPLICATION TO MULTIPLICATIVE ANALOGUE OF C-CLASS FUNCTIONS, JP Journal of Fixed Point Theory and Applications, August 2016, Volume 11, Issue 2, Pages 113 - 124

(6) Jeffery Ezearn, Higher-order Lipschitz Mappings, Fixed Point Theory and Applications (2015) 2015:88

(7) Clement Ampadu(2015). Generalization of Higher Order Contraction Mapping Theorem. Unpublished

(8) S. Sadiq Basha and N. Shahzad, Best proximity point theorems for generalized proximal contractions, Fixed Point Theory Appl., 2012(2012), Article ID 42

(9) AJAY SHARMA, BALWANT SINGH THAKUR, BEST PROXIMITY POINTS FOR K-PROXIMAL CONTRACTION, International Journal of Analysis and Applications, Volume 6, Number 1 (2014), 82-88

(10) Clement Ampadu, Characterization Theorems Inspired by the Hardy-Rogers Map I: Some Results in Metric Spaces. lulu.com, 2016. ISBN: 1365101185, 9781365101182

(11) S.K. Chatterjee, Fixed point theorems, Comptes. Rend. Acad. Bulgaria Sci. 25(1972), 727-730

Chapter 3

Best Proximity Point Theorems in Multiplicative S-Metric Space

3.1 Brief Summary

> **Abstract C.1 1**
>
> The notion of S-metric [S. Sedghi, N. Shobe, A. Aliouche, A generalization of fixed point theorems in S-metric spaces, Mat. Vesnik, 64(2012), 258-266] is a modification of D^*-metric [S. Sedghi, N. Shobe, H. Zhou, A common fixed point theorem in D^*-metric spaces, Fixed Point Theory Appl., 2007 (2007), Article ID 27906, 1-13] and G-metric [Z. Mustafa, B. Sims, A new approach to generalized metric spaces, J. Nonlinear Convex Anal., 7 (2006), no. 2, 289-297]. In this chapter we introduce multiplicative version of S-metric and obtain some best proximity point results for a class of proximal multiplicative contractive mappings defined implicitly using the C-class function [Clement Ampadu and Arslan Hojat Ansari, FIXED POINT THEOREMS IN COMPLETE MULTIPLICATIVE METRIC SPACES WITH APPLICATION TO MULTIPLICATIVE ANALOGUE OF C-CLASS FUNCTIONS, JP Journal of Fixed Point Theory and Applications, August 2016, Volume 11, Issue 2, Pages 113 - 124]

3.2 Preliminaries

> **Definition C.1 1**
>
> Let A and B be two nonempty subsets of a multiplicative metric space (X, m). We say $x \in A$ is a fixed point of $T : A \mapsto B$ if $Tx = x$

> **Definition C.2 1**
>
> An element $a \in A$ will be called a best proximity point of T if $m(a, Ta) = m(A, B)$, where $m(A, B) = \inf\{m(x, y) : x \in A, y \in B\}$

From [S. Sedghi, N. Shobe, A. Aliouche, A generalization of fixed point theorems in S-metric spaces, Mat. Vesnik, 64(2012), 258-266] we deduce the following

Definition C.3 1

Let X be a nonempty set. A multiplicative S-metric on X will be a function $S_m : X^3 \mapsto [1, \infty)$ such that for each $x, y, z, a \in X$, the following holds

(a) $S_m(x, y, z) \geq 1$

(b) $S_m(x, y, z) = 1$ iff $x = y = z$

(c) $S_m(x, y, z) \leq S_m(x, x, a) \cdot S_m(y, y, a) \cdot S_m(z, z, a)$

We will write (X, S_m) to denote a multiplicative S-metric space

Remark C.4 1

The above notion is a multiplicative generalization of D^*-metric [S. Sedghi, N. Shobe, H. Zhou, A common fixed point theorem in D^*-metric spaces, Fixed Point Theory Appl., 2007 (2007), Article ID 27906, 1-13] and G-metric [Z. Mustafa, B. Sims, A new approach to generalized metric spaces, J. Nonlinear Convex Anal., 7 (2006), no. 2, 289-297].

Remark C.5 1

Note that every multiplicative S-metric on X induces a multiplicative metric m_{S_m} on X defined by $m_{S_m}(x, y) = S_m(x, x, y) \cdot S_m(y, y, x)$ for all $x, y \in X$

Taking inspiration from [S. Sedghi, N. Shobe, A. Aliouche, A generalization of fixed point theorems in S-metric spaces, Mat. Vesnik 64(2012), 258-266] we have the following

Example C.6 1

Let $X = \mathbb{R}^2$ and m be an ordinary multiplicative metric on X. Put

$$S_m(x, y, z) = m(x, y) \cdot m(x, z) \cdot m(y, z)$$

for all $x, y, z \in \mathbb{R}$. Then S_m is a multiplicative S-metric space on X

Taking inspiration from [S. Sedghi, N. Shobe, A. Aliouche, A generalization of fixed point theorems in S-metric spaces, Mat. Vesnik 64(2012), 258-266] we have the following

Lemma C.7 1

Let (X, S_m) be a multiplicative S-metric space. Then $S_m(x, x, y) = S_m(y, y, x)$ for all $x, y \in X$

Taking inspiration from [N.T. Hieu, N.T. Thanh Ly, N.V. Dung, A generalization of Ciric quasi-contractions for maps on S-metric spaces, Thai J. Math., 13 (2015), no.2, 369-380] we have the following

Lemma C.8 1

Let (X, S_m) be a multiplicative S-metric space. Then

$$S_m(x, x, z) \leq S_m(x, x, y)^2 \cdot S_m(y, y, z)$$

and

$$S_m(x, x, z) \leq S_m(x, x, y)^2 \cdot S_m(z, z, y)$$

for all $x, y, z \in X$

Taking inspiration from [S. Sedghi, N. Shobe, A. Aliouche, A generalization of fixed point theorems in S-metric spaces, Mat. Vesnik 64(2012), 258-266] we have the following

CHAPTER 3. BEST PROXIMITY POINT THEOREMS IN MULTIPLICATIVE S-METRIC SPACE

Definition C.9 1

Let (X, S_m) be a multiplicative S-metric space

(a) A sequence $\{x_n\}$ in X is said to converge to $x \in X$ if $S_m(x_n, x_n, x) \to 1$ as $n \to \infty$. That is, for each $\epsilon > 1$, there exists $N \in \mathbb{N}$ such that for all $n \geq N$ we have $S_m(x_n, x_n, x) < \epsilon$

(b) A sequence $\{x_n\}$ in X is called Cauchy if $S_m(x_n, x_n, x_k) \to 1$ as $n, k \to \infty$. That is, for each $\epsilon > 1$, there exists $N \in \mathbb{N}$ such that for all $n, k \geq N$ we have $S_m(x_n, x_n, x_k) < \epsilon$

(c) The multiplicative S-metric space (X, S_m) is said to be complete if every Cauchy sequence in X converges to a point in X

Taking inspiration from [S. Sedghi, N. Shobe, A. Aliouche, A generalization of fixed point theorems in S-metric spaces, Mat. Vesnik 64(2012), 258-266] we have the following

Lemma C.10 1

Let (X, S_m) be a multiplicative S-metric space

(a) If $x_n \to x$ and $y_n \to y$, then $S_m(x_n, x_n, y_n) \to S_m(x, x, y)$

(b) If the sequence $\{x_n\}$ in X converges to x, then x is unique

(c) If the sequence $\{x_n\}$ in X converges to x, then $\{x_n\}$ is Cauchy

Notation C.11 1

Φ will denote the class of all functions $\phi : [1, \infty) \mapsto [1, \infty)$ which satisfy

(a) ϕ is continuous and nondecreasing

(b) $\phi(t) = 1$ iff $t = 1$

(c) $\phi(t + s) \leq \phi(t) \cdot \phi(t)$, for all $t, s \in [1, \infty)$

Notation C.12 1

Ψ will denote the class of all functions $\psi : [1, \infty) \mapsto [1, \infty)$ such that ψ is lower semi-continuous, and $\psi(t) = 1$ iff $t = 1$

Notation C.13 1

Let A and B be two nonempty subsets of a multiplicative S-metric space (X, S_m). A_0 will denote the set $\{x \in A : m_{S_m}(x, y) = m_{S_m}(A, B)\ for\ some\ y \in B\}$ and B_0 will denote the set $\{x \in B : m_{S_m}(x, y) = m_{S_m}(A, B)\ for\ some\ y \in A\}$, where $m_{S_m}(A, B) = \inf\{m_{S_m}(x, y) : x \in A, y \in B\}$, and $m_{S_m}(x, y) = S_m(x, x, y) \cdot S_m(y, y, x)$

Definition C.14 1

Let (X, S_m) be a multiplicative S-metric space, and let A and B be two nonempty subsets of X. Then B is said to be approximately compact with respect to A if every sequence $\{y_n\}$ in B, satisfying the condition $m_{S_m}(x, y_n) \to m_{S_m}(x, B)$ for some $x \in A$, has a convergent subsequence

> **Definition C.15 1**
>
> Let A and B be two nonempty subsets of a multiplicative S-metric space (X, S_m). Let $T : A \mapsto B$ be a non-self mapping. We will say T is an implicit $S-\phi-\psi$ proximal multiplicative contraction, if for $x, y, u, v \in A$, $m_{S_m}(u, Tx) = m_{S_m}(A, B)$ and $m_{S_m}(v, Ty) = m_{S_m}(A, B)$ implies
>
> $$\phi(S_m(u,u,v)) \leq F\left[\frac{\phi(S_m(x,x,y))}{\psi(S_m(x,x,y))}\right]$$
>
> where $F(x,y) := F\left(\frac{x}{y}\right)$ is a multiplicative c-class function [Clement Ampadu and Arslan Hojat Ansari, FIXED POINT THEOREMS IN COMPLETE MULTIPLICATIVE METRIC SPACES WITH APPLICATION TO MULTIPLICATIVE ANALOGUE OF C-CLASS FUNCTIONS, JP Journal of Fixed Point Theory and Applications, August 2016, Volume 11, Issue 2, Pages 113 - 124], $\phi \in \Phi$, and $\psi \in \Psi$

3.3 Main Results

> **Theorem C.1 1**
>
> Let A and B be two nonempty subsets of a multiplicative S-metric space (X, S_m) such that (A, S_m) is a complete multiplicative S-metric space, A_0 is nonempty, and B is approximately compact with respect to A. Assume that $T : A \mapsto B$ is an implicit $S - \phi - \psi$ proximal multiplicative contractive mapping such that $T(A_0) \subseteq B_0$. Then T has a unique best proximity point, that is, there is a unique $z \in A$ such that $m_{S_m}(z, Tz) = m_{S_m}(A, B)$.

> **Proof of Theorem C.1 1**
>
> Since the subset A_0 is not empty, we take x_0 in A_0. Taking $Tx_0 \in T(A_0) \subseteq B_0$ into account, we can find $x_1 \in A_0$ such that $m_{S_m}(x_1, Tx_0) = m_{S_m}(A, B)$. Further, since $Tx_1 \in T(A_0) \subseteq B_0$, it follows that there exists an element x_2 in A_0 such that $m_{S_m}(x_2, Tx_1) = m_{S_m}(A, B)$. Continuing, we obtain for all $n \in \mathbb{N} \cup \{0\}$ a sequence $\{x_n\}$ in A_0 satisfying $m_{S_m}(x_{n+1}, Tx_n) = m_{S_m}(A, B)$. It follows that $m_{S_m}(u, Tx) = m_{S_m}(A, B)$ and $m_{S_m}(v, Ty) = m_{S_m}(A, B)$, where $u = x_n, x = x_{n-1}, v = x_{n+1}, y = x_n$. Therefore from the contractive definition of the theorem we have
>
> $$\phi(S_m(x_n, x_n, x_{n+1})) \leq F\left[\frac{\phi(S_m(x_{n-1}, x_{n-1}, x_n))}{\psi(S_m(x_{n-1}, x_{n-1}, x_n))}\right]$$
>
> which by the properties of F and ϕ implies $S_m(x_n, x_n, x_{n+1}) \leq S_m(x_{n-1}, x_{n-1}, x_n)$. It follows that the sequence $\{S_m(x_n, x_n, x_{n+1})\}$ is a decreasing sequence in $[1, \infty)$ and thus converges to some $t \in [1, \infty)$. We claim that $t = 1$, if not, then taking limits in
>
> $$\phi(S_m(x_n, x_n, x_{n+1})) \leq F\left[\frac{\phi(S_m(x_{n-1}, x_{n-1}, x_n))}{\psi(S_m(x_{n-1}, x_{n-1}, x_n))}\right]$$
>
> we deduce that $\phi(t) \leq F\left[\frac{\phi(t)}{\psi(t)}\right]$ which is a contradiction, thus $t = 1$, that is,
>
> $$\lim_{n \to \infty} S_m(x_n, x_n, x_{n+1}) = 1$$
>
> Now we show that $\{x_n\}$ is a multiplicative Cauchy sequence. Suppose not, then there exists $\epsilon > 1$ and a subsequence $\{x_{n_k}\}$ of $\{x_n\}$ such that $S_m(x_{v_k}, x_{v_k}, x_{n_k}) \geq \epsilon$ with $n_k \geq v_k > k$. Further corresponding to v_k we can choose n_k in such a way that it is the smallest integer with $n_k > v_k$ and satisfying $S_m(x_{v_k}, x_{v_k}, x_{n_k}) \geq \epsilon$. Hence, $S_m(x_{v_k}, x_{v_k}, x_{n_k-1}) < \epsilon$. Set $\beta_n = S_m(x_n, x_n, x_{n-1})^2$. By Lemma C.7 and Lemma C.8, we deduce that
>
> $$\epsilon \leq S_m(x_{v_k}, x_{v_k}, x_{n_k})$$
> $$= S_m(x_{n_k}, x_{n_k}, x_{n_k-1})^2 \cdot S_m(x_{v_k}, x_{v_k}, x_{n_k-1})$$
> $$\leq S_m(x_{n_k}, x_{n_k}, x_{n_k-1})^2 \cdot \epsilon$$
> $$\leq \beta_{n_k} \cdot \epsilon$$
>
> Now taking limits in the above we deduce that $\lim_{k \to \infty} S_m(x_{v_k}, x_{v_k}, x_{n_k}) = \epsilon$. Now by Lemma C.8, we have the following two inequalities
>
> $$S_m(x_{v_k}, x_{v_k}, x_{n_k}) \leq S_m(x_{v_k}, x_{v_k}, x_{v_k-1})^2 \cdot S_m(x_{n_k}, x_{n_k}, x_{v_k-1})$$
> $$\leq S_m(x_{v_k}, x_{v_k}, x_{v_k-1})^2 \cdot S_m(x_{n_k}, x_{n_k}, x_{n_k-1})^2 \cdot S_m(x_{v_k-1}, x_{v_k-1}, x_{n_k-1})$$
> $$= (\beta_{v_k} \cdot \beta_{n_k}) \cdot S_m(x_{v_k-1}, x_{v_k-1}, x_{n_k-1})$$
>
> and
>
> $$S_m(x_{v_k-1}, x_{v_k-1}, x_{n_k-1}) \leq S_m(x_{v_k-1}, x_{v_k-1}, x_{v_k})^2 \cdot S_m(x_{n_k-1}, x_{n_k-1}, x_{v_k})$$
> $$\leq S_m(x_{v_k-1}, x_{v_k-1}, x_{v_k})^2 \cdot S_m(x_{n_k-1}, x_{n_k-1}, x_{n_k})^2 \cdot S_m(x_{v_k}, x_{v_k}, x_{n_k})$$
> $$= (\beta_{v_k-1} \cdot \beta_{n_k-1}) \cdot S_m(x_{v_k}, x_{v_k}, x_{n_k})$$
>
> Now taking limits as $k \to \infty$ in the second of the two inequalities immediately above, and applying the first of the two inequalities immediately above, we deduce that
>
> $$\lim_{k \to \infty} S_m(x_{v_k-1}, x_{v_k-1}, x_{n_k-1}) = \epsilon$$
>
> Now from the contractive definition of the theorem we deduce that
>
> $$\phi(S_m(x_{v_k}, x_{v_k}, x_{n_k})) \leq F\left[\frac{\phi(S_m(x_{v_k-1}, x_{v_k-1}, x_{n_k-1}))}{\psi(S_m(x_{v_k-1}, x_{v_k-1}, x_{n_k-1}))}\right]$$

Proof of Theorem C.1 continued 1

Taking limits in the above inequality as $k \to \infty$, we deduce that $\phi(\epsilon) \leq F\left[\frac{\phi(\epsilon)}{\psi(\epsilon)}\right]$, and clearly we have equality iff $\psi(\epsilon) = 1$, which is a contradiction since it would imply $\epsilon = 1$, however by assumption $\epsilon > 1$. It follows that, $\lim_{n,k \to \infty} S_m(x_k, x_k, x_n) = 1$, thus, $\{x_n\}$ is multiplicative Cauchy. Since (A, S_m) is a complete multiplicative S-metric space, there exists $z \in A$ such that $\lim_{n \to \infty} x_n = z$. Now observe for all $n \in \mathbb{N}$, we have

$$m_{S_m}(z, B) \leq m_{S_m}(z, Tx_n)$$
$$\leq m_{S_m}(z, x_{n+1}) \cdot m_{S_m}(x_{n+1}, Tx_n)$$
$$= m_{S_m}(z, x_{n+1}) \cdot m_{S_m}(A, B)$$

Now taking limits in the above inequality as $n \to \infty$ we deduce that

$$\lim_{n \to \infty} m_{S_m}(z, Tx_n) = m_{S_m}(z, B) = m_{S_m}(A, B)$$

Since B is approximately compact with respect to A, it follows that $\{Tx_n\}$ has a subsequence $\{Tx_{n_k}\}$ that converges to some $y^* \in B$. Hence

$$m_{S_m}(z, y^*) = \lim_{n \to \infty} m_{S_m}(x_{n_k+1}, Tx_{n_k}) = m_{S_m}(A, B)$$

and so $z \in A_0$. Now since $Tz \in T(A_0) \subseteq B_0$, there exists $w \in A_0$ such that $m_{S_m}(w, Tz) = m_{S_m}(A, B)$. From the contractive definition of the theorem we deduce that

$$\phi(S_m(x_{n+1}, x_{n+1}, w)) \leq F\left[\frac{\phi(S_m(x_n, x_n, z))}{\psi(S_m(x_n, x_n, z))}\right]$$

Taking limits in the above as $n \to \infty$, we deduce that $\phi(S_m(z, z, w)) \leq F\left[\frac{\phi(1)}{\psi(1)}\right]$ which implies $S_m(z, z, w) = 1$, that is, $z = w$. Thus, $m_{S_m}(z, Tz) = m_{S_m}(A, B)$ and T has best proximity point. Finally, we show uniqueness of the best proximity point. Suppose $p \neq q$ is such that $m_{S_m}(p, Tp) = m_{S_m}(A, B)$ and $m_{S_m}(q, Tq) = m_{S_m}(A, B)$. By the contractive definition of the theorem, we deduce that

$$\phi(S_m(p, p, q)) \leq F\left[\frac{\phi(S_m(p, p, q))}{\psi(S_m(p, p, q))}\right]$$

which implies $\psi(S_m(p, p, q)) = 1$, that is, $S_m(p, p, q) = 1$ and hence $p = q$

If in the previous theorem, we take $\phi(t) = t$, $\psi(t) = t^{1-r}$, where $0 \leq r < 1$, and let the multiplicative C-class function be given by $F(x, y) := F\left(\frac{x}{y}\right) = \frac{x}{y}$, then we get the following

Corollary C.2 1

Let A and B be two nonempty subsets of a multiplicative S-metric space (X, S_m) such that (A, S_m) is a complete multiplicative S-metric space, A_0 is nonempty, and B is approximately compact with respect to A. Assume that $T : A \mapsto B$ is a non-self mapping such that $T(A_0) \subseteq B_0$ and for $x, y, u, v \in A$, $m_{S_m}(u, Tx) = m_{S_m}(A, B)$ and $m_{S_m}(v, Ty) = m_{S_m}(A, B)$ implies $S_m(u, u, v) \leq S_m(x, x, y)^r$, where $0 \leq r < 1$. Then T has a unique best proximity point, that is, there is a unique $z \in A$ such that $m_{S_m}(z, Tz) = m_{S_m}(A, B)$

3.4 Exercises

Exercise C.1 1

Taking inspiration from [J. Nantadilok, Best Proximity Point Results in S-metric Spaces, International Journal of Mathematical Analysis Vol. 10, 2016, no. 27, 1333 - 1346] give an example in support of Theorem C.1

Exercise C.2 1

Let A and B be two nonempty subsets of a multiplicative S-metric space (X, S_m). Let $T : A \mapsto B$ be a non-self mapping. We will say T is an implicit $S - \phi - \psi$ proximal multiplicative Reich type contraction, if for $x, y, u, v \in A$, $m_{S_m}(u, Tx) = m_{S_m}(A, B)$ and $m_{S_m}(v, Ty) = m_{S_m}(A, B)$ implies

$$\phi(S_m(u,u,v)) \leq F\left[\frac{\phi(S_m(x,x,y) \cdot S_m(u,u,x) \cdot S_m(v,v,y))}{\psi(S_m(x,x,y) \cdot S_m(u,u,x) \cdot S_m(v,v,y))}\right]$$

where $F(x,y) := F\left(\frac{x}{y}\right)$ is a multiplicative c-class function [Clement Ampadu and Arslan Hojat Ansari, FIXED POINT THEOREMS IN COMPLETE MULTIPLICATIVE METRIC SPACES WITH APPLICATION TO MULTIPLICATIVE ANALOGUE OF C-CLASS FUNCTIONS, JP Journal of Fixed Point Theory and Applications, August 2016, Volume 11, Issue 2, Pages 113 - 124], $\phi \in \Phi$, and $\psi \in \Psi$

(a) Let A and B be two nonempty subsets of a multiplicative S-metric space (X, S_m) such that (A, S_m) is a complete multiplicative S-metric space, A_0 is nonempty, and B is approximately compact with respect to A. Assume that $T : A \mapsto B$ is an implicit $S - \phi - \psi$ proximal multiplicative Reich type contractive mapping such that $T(A_0) \subseteq B_0$. Show T has a unique best proximity point, that is, there is a unique $z \in A$ such that $m_{S_m}(z, Tz) = m_{S_m}(A, B)$

(b) State the Corollary arising from (a), if we take $\phi(t) = t$, $\psi(t) = t^{1-r}$, where $0 \leq r < \frac{1}{3}$, and let the multiplicative C-class function be given by $F(x,y) := F\left(\frac{x}{y}\right) = \frac{x}{y}$

Exercise C.3 1

Let A and B be two nonempty subsets of a multiplicative S-metric space (X, S_m). Let $T : A \mapsto B$ be a non-self mapping. We will say T is an implicit $S - \phi - \psi$ proximal multiplicative Kannan type contraction, if for $x, y, u, v \in A$, $m_{S_m}(u, Tx) = m_{S_m}(A, B)$ and $m_{S_m}(v, Ty) = m_{S_m}(A, B)$ implies

$$\phi(S_m(u,u,v)) \leq F\left[\frac{\phi(S_m(u,u,x) \cdot S_m(v,v,y))}{\psi(S_m(u,u,x) \cdot S_m(v,v,y))}\right]$$

where $F(x,y) := F\left(\frac{x}{y}\right)$ is a multiplicative c-class function [Clement Ampadu and Arslan Hojat Ansari, FIXED POINT THEOREMS IN COMPLETE MULTIPLICATIVE METRIC SPACES WITH APPLICATION TO MULTIPLICATIVE ANALOGUE OF C-CLASS FUNCTIONS, JP Journal of Fixed Point Theory and Applications, August 2016, Volume 11, Issue 2, Pages 113 - 124], $\phi \in \Phi$, and $\psi \in \Psi$

(a) Let A and B be two nonempty subsets of a multiplicative S-metric space (X, S_m) such that (A, S_m) is a complete multiplicative S-metric space, A_0 is nonempty, and B is approximately compact with respect to A. Assume that $T : A \mapsto B$ is an implicit $S - \phi - \psi$ proximal multiplicative Kannan type contractive mapping such that $T(A_0) \subseteq B_0$. Show T has a unique best proximity point, that is, there is a unique $z \in A$ such that $m_{S_m}(z, Tz) = m_{S_m}(A, B)$

(b) State the Corollary arising from (a), if we take $\phi(t) = t$, $\psi(t) = t^{1-r}$, where $0 \leq r < \frac{1}{2}$, and let the multiplicative C-class function be given by $F(x,y) := F\left(\frac{x}{y}\right) = \frac{x}{y}$

> **Exercise C.4 1**
>
> Let A and B be two nonempty subsets of a multiplicative S-metric space (X, S_m). Let $T : A \mapsto B$ be a non-self mapping. We will say T is an implicit $S-\phi-\psi$ proximal multiplicative Hardy and Rogers type contraction, if for $x, y, u, v \in A$, $m_{S_m}(u, Tx) = m_{S_m}(A, B)$ and $m_{S_m}(v, Ty) = m_{S_m}(A, B)$ implies
>
> $$\phi(S_m(u,u,v)) \leq F\left[\frac{\phi(S_m(x,x,y) \cdot S_m(u,u,x) \cdot S_m(u,u,y) \cdot S_m(v,v,x) \cdot S_m(v,v,y))}{\psi(S_m(x,x,y) \cdot S_m(u,u,x) \cdot S_m(u,u,y) \cdot S_m(v,v,x) \cdot S_m(v,v,y))}\right]$$
>
> where $F(x,y) := F\left(\frac{x}{y}\right)$ is a multiplicative c-class function [Clement Ampadu and Arslan Hojat Ansari, FIXED POINT THEOREMS IN COMPLETE MULTIPLICATIVE METRIC SPACES WITH APPLICATION TO MULTIPLICATIVE ANALOGUE OF C-CLASS FUNCTIONS, JP Journal of Fixed Point Theory and Applications, August 2016, Volume 11, Issue 2, Pages 113 - 124], $\phi \in \Phi$, and $\psi \in \Psi$
>
> (a) Let A and B be two nonempty subsets of a multiplicative S-metric space (X, S_m) such that (A, S_m) is a complete multiplicative S-metric space, A_0 is nonempty, and B is approximately compact with respect to A. Assume that $T : A \mapsto B$ is an implicit $S - \phi - \psi$ proximal multiplicative Hardy and Rogers type contractive mapping such that $T(A_0) \subseteq B_0$. Show T has a unique best proximity point, that is, there is a unique $z \in A$ such that $m_{S_m}(z, Tz) = m_{S_m}(A, B)$
>
> (b) State the Corollary arising from (a), if we take $\phi(t) = t$, $\psi(t) = t^{1-r}$, where $0 \leq r < \frac{1}{5}$, and let the multiplicative C-class function be given by $F(x,y) := F\left(\frac{x}{y}\right) = \frac{x}{y}$

3.5 References

(1) S. Sedghi, N. Shobe, A. Aliouche, A generalization of fixed point theorems in S-metric spaces, Mat. Vesnik, 64(2012), 258-266

(2) S. Sedghi, N. Shobe, H. Zhou, A common fixed point theorem in D^*-metric spaces, Fixed Point Theory Appl., 2007 (2007), Article ID 27906, 1-13

(3) Z. Mustafa, B. Sims, A new approach to generalized metric spaces, J. Nonlinear Convex Anal., 7 (2006), no. 2, 289-297

(4) Clement Ampadu and Arslan Hojat Ansari, FIXED POINT THEOREMS IN COMPLETE MULTIPLICATIVE METRIC SPACES WITH APPLICATION TO MULTIPLICATIVE ANALOGUE OF C-CLASS FUNCTIONS, JP Journal of Fixed Point Theory and Applications, August 2016, Volume 11, Issue 2, Pages 113 - 124

(5) N.T. Hieu, N.T. Thanh Ly, N.V. Dung, A generalization of Ciric quasi-contractions for maps on S-metric spaces, Thai J. Math., 13 (2015), no.2, 369-380

(6) J. Nantadilok, Best Proximity Point Results in S-metric Spaces, International Journal of Mathematical Analysis Vol. 10, 2016, no. 27, 1333 -1346

Chapter 4

Best Proximity Point Theorems for Implicit Generalized Proximal C-Multiplicative Contraction with Partial Orders

4.1 Brief Summary

> **Abstract D.1 1**
>
> The notion of weakly C-contraction was introduced in [Choudhury, BS: Unique fixed point theorem for weak-contractive mappings. Kathmandu Univ. J. Sci. Eng. Technol. 5, 6-13 (2009)]. In this chapter we introduce implicit weakly C-multiplicative contraction to the case of non-self mappings and obtain some best proximity point theorems for this class.

4.2 Preliminaries

Taking inspiration from [Chatterjea, SK: Fixed point theorems. C. R. Acad. Bulgare Sci. 25, 727-730 (1972)] we introduce the following

> **Definition D.1 1**
>
> Let (X, m) be a multiplicative metric space. A map $T : X \mapsto X$ will be called a C-multiplicative contraction if there exists $\alpha \in \left(0, \frac{1}{2}\right)$ such that for all $x, y \in X$, $m(Tx, Ty) \leq [m(x, Ty) \cdot m(y, Tx)]^\alpha$

Taking inspiration from [Choudhury, BS: Unique fixed point theorem for weak-contractive mappings. Kathmandu Univ. J. Sci. Eng. Technol. 5, 6-13 (2009)] we introduce the following

CHAPTER 4. BEST PROXIMITY POINT THEOREMS FOR IMPLICIT GENERALIZED PROXIMAL C-MULTIPLICATIVE CONTRACTION WITH PARTIAL ORDERS

Definition D.2 1

Let (X, m) be a multiplicative metric space. A mapping $T : X \mapsto X$ will be called an implicit weakly C- multiplicative contraction if for all $x, y \in X$, we have

$$m(Tx, Ty) \leq F\left[\frac{\sqrt{m(x, Ty) \cdot m(y, Tx)}}{\psi(m(x, Ty), m(y, Tx))}\right]$$

where $\psi : [1, \infty)^2 \mapsto [1, \infty)$ is a continuous and nondecreasing function such that $\psi(x, y) = 1$ iff $x = y = 1$, and $F(x, y) := F\left(\frac{x}{y}\right)$ is a multiplicative C-class function [Clement Ampadu and Arslan Hojat Ansari, FIXED POINT THEOREMS IN COMPLETE MULTIPLICATIVE METRIC SPACES WITH APPLICATION TO MULTIPLICATIVE ANALOGUE OF C-CLASS FUNCTIONS, JP Journal of Fixed Point Theory and Applications, August 2016, Volume 11, Issue 2, Pages 113-124]

Notation D.3 1

Let A and B be nonempty subset of a multiplicative metric space (X, m).

$$m(A, B) := \inf\{m(x, y) : x \in A, y \in B\}$$

$$A_0 := \{x \in A : m(x, y) = m(A, B) \text{ for some } y \in B\}$$

$$B_0 := \{y \in B : m(x, y) = m(A, B) \text{ for some } x \in A\}$$

Notation D.4 1

$(X, \preceq)$ will denote a partially ordered set

Definition D.5 1

A mapping $T : A \mapsto B$ is said to be increasing if $x \preceq y$ implies $Tx \preceq Ty$ for all $x, y \in A$

Taking inspiration from [Sadiq Basha, S: Best proximity point theorems on partially ordered sets. Optim. Lett. (2012). doi:10.1007/s11590-012-0489-1] we introduce the following in multiplicative metric spaces

Definition D.6 1

A mapping $T : A \mapsto B$ will be called proximally order-preserving iff it satisfies $x \preceq y$, $m(u, Tx) = m(A, B)$, and $m(v, Ty) = m(A, B)$ implies $u \preceq v$ for all $u, v, x, y \in A$

Remark D.7 1

For a self-mapping Definition D.6 reduces to Definition D.5

Definition D.8 1

Let (X, m) be a multiplicative metric space. A point $x \in A$ will be called a best proximity point of the mapping $T : A \mapsto B$ if $m(x, Tx) = m(A, B)$

Chapter 4. Best Proximity Point Theorems for Implicit Generalized Proximal C-Multiplicative Contraction with Partial Orders

Definition D.9 1

A mapping $T : A \mapsto B$ will be called an implicit generalized proximal C-multiplicative contraction, if for all $u, v, x, y \in A$, it satisfies

$$x \preceq y, m(u, Tx) = m(A, B), m(v, Ty) = m(A, B)$$

implies

$$m(u, v) \leq F\left[\frac{\sqrt{m(x,v) \cdot m(y,u)}}{\psi(m(x,v), m(y,u))}\right]$$

where $\psi : [1, \infty)^2 \mapsto [1, \infty)$ is a continuous and nondecreasing function such that $\psi(x, y) = 1$ iff $x = y = 1$, and $F(x, y) := F\left(\frac{x}{y}\right)$ is a multiplicative C-class function [Clement Ampadu and Arslan Hojat Ansari, FIXED POINT THEOREMS IN COMPLETE MULTIPLICATIVE METRIC SPACES WITH APPLICATION TO MULTIPLICATIVE ANALOGUE OF C-CLASS FUNCTIONS, JP Journal of Fixed Point Theory and Applications, August 2016, Volume 11, Issue 2, Pages 113-124]

Remark D.10 1

For a self-mapping Definition D.9 reduces to Definition D.2

4.3 Main Results

Theorem D.1 1

Let X be a nonempty set such that $(X, \preceq)$ is a partially ordered set and let (X, m) be a complete multiplicative metric space. Let A and B be nonempty closed subsets of X such that A_0 and B_0 are nonempty. Let $T : A \mapsto B$ satisfy the following conditions

(a) T is a continuous, proximally order-preserving and implicit generalized proximal C-multiplicative contraction such that $T(A_0) \subseteq B_0$

(b) there exists an element x_0 and x_1 in A_0 such that $x_0 \preceq x_1$ and $m(x_1, Tx_0) = m(A, B)$

Then there exists a point $x \in A$ such that $m(x, Tx) = m(A, B)$. Moreover, for any fixed $x_0 \in A_0$, the sequence $\{x_n\}$ defined by $m(x_{n+1}, Tx_n) = m(A, B)$ converges to the point x

Proof of Theorem D.1 1

By (b) of the Theorem, there exists $x_0, x_1 \in A_0$ such that $x_0 \preceq x_1$ and $m(x_1, Tx_0) = m(A, B)$. Since $T(A_0) \subseteq B_0$ there exists a point $x_2 \in A_0$ such that $m(x_2, Tx_1) = m(A, B)$. By proximally order-preserving property of T, we get $x_1 \preceq x_2$. Continuing this process we can find a sequence $\{x_n\}$ in A_0 such that $x_{n-1} \preceq x_n$ and $m(x_n, Tx_{n-1}) = m(A, B)$. Having found the point x_n, one can choose a point $x_{n+1} \in A_0$ such that $x_n \preceq x_{n+1}$ and $m(x_{n+1}, Tx_n) = m(A, B)$. By the contractive condition of the theorem and the properties of F we have

$$m(x_n, x_{n+1}) \leq F\left[\frac{\sqrt{m(x_{n-1}, x_{n+1}) \cdot m(x_n, x_n)}}{\psi(m(x_{n-1}, x_{n+1}), m(x_n, x_n))}\right]$$
$$= F\left[\frac{\sqrt{m(x_{n-1}, x_{n+1})}}{\psi(m(x_{n-1}, x_{n+1}), 1)}\right]$$
$$\leq \sqrt{m(x_{n-1}, x_{n+1})}$$
$$\leq \sqrt{m(x_{n-1}, x_n) \cdot m(x_n, x_{n+1})}$$

From the above it follows that $m(x_n, x_{n+1}) \leq m(x_{n-1}, x_n)$, that is, the sequence $\{m(x_n, x_{n+1})\}$ is nonincreasing and bounded below. It follows that there exists $r \geq 1$ such that

$$\lim_{n \to \infty} m(x_{n+1}, x_n) = r$$

If we take limits as $n \to \infty$ in the chain of inequalities immediately above, we deduce that

$$r \leq \lim_{n \to \infty} \sqrt{m(x_{n-1}, x_{n+1})} \leq \sqrt{r \cdot r} = r$$

From which it follows that $\lim_{n \to \infty} m(x_{n-1}, x_{n+1}) = r^2$. Again taking limits as $n \to \infty$ in the chain of inequalities immediately above, and using the fact that

$$\lim_{n \to \infty} m(x_{n-1}, x_{n+1}) = r^2$$

and $\lim_{n \to \infty} m(x_{n+1}, x_n) = r$, the continuity of ψ and properties of F, we deduce that

$$r \leq \sqrt{r^2} = F\left[\frac{r}{\psi(r^2, 1)}\right] \leq r$$

It follows that $\psi(r^2, 1) = 1$ and hence $r = 1$, that is, $\lim_{n \to \infty} m(x_{n+1}, x_n) = 1$. Now we show that $\{x_n\}$ is a multiplicative Cauchy sequence. Suppose $\{x_n\}$ is not a multiplicative Cauchy sequence, then there exists $\epsilon > 1$ and subsequences $\{x_{v_k}\}, \{x_{n_k}\}$ of $\{x_n\}$ such that $n_k > m_k \geq k$ with $r_k := m(x_{v_k}, x_{n_k}) \geq \epsilon$ and $m(x_{v_k}, x_{n_k-1}) < \epsilon$ for each $k \in \mathbb{N}$. For each $n \in \mathbb{N}$, let $\alpha_n := m(x_{n+1}, x_n)$. Now observe that

$$\epsilon \leq r_k$$
$$\leq m(x_{v_k}, x_{n_k-1}) \cdot m(x_{n_k-1}, x_{n_k})$$
$$< \epsilon \cdot \alpha_{n_k-1}$$

Since $\lim_{n \to \infty} m(x_{n+1}, x_n) = 1$, from the above we deduce that $\lim_{k \to \infty} r_k = \epsilon$. Now observe that

$$r_k = m(x_{n_k}, x_{v_k})$$
$$\leq m(x_{n_k}, x_{v_k+1}) \cdot m(x_{v_k+1}, x_{v_k})$$
$$= m(x_{n_k}, x_{v_k+1}) \cdot \alpha_{v_k}$$
$$\leq m(x_{n_k}, x_{v_k}) \cdot m(x_{v_k}, x_{v_k+1}) \cdot \alpha_{v_k}$$
$$= r_k \cdot \alpha_{v_k}^2$$

CHAPTER 4. BEST PROXIMITY POINT THEOREMS FOR IMPLICIT GENERALIZED PROXIMAL C-MULTIPLICATIVE CONTRACTION WITH PARTIAL ORDERS

Proof of Theorem D.1 Continued 1

Since $\lim_{n\to\infty} m(x_{n+1}, x_n) = 1$ and $\lim_{k\to\infty} r_k = \epsilon$, if we take limits as $k \to \infty$ in the above we get $\lim_{n\to\infty} m(x_{n_k}, x_{v_k+1}) = \epsilon$. Similarly, we have $\lim_{k\to\infty} m(x_{v_k}, x_{n_k+1}) = \epsilon$. On the other hand, by the construction of $\{x_n\}$, we may assume that $x_{v_k} \preceq x_{n_k}$ such that

$$m(x_{n_k+1}, Tx_{n_k}) = m(A, B) \text{ and } m(x_{v_k+1}, Tx_{v_k}) = m(A, B)$$

From the equality statements immediately above, the multiplicative triangle inequality, and the contractive condition of the theorem, we deduce the following

$$\epsilon \leq r_k$$
$$\leq m(x_{v_k}, x_{v_k+1}) \cdot m(x_{n_k+1}, x_{n_k}) \cdot m(x_{v_k+1}, x_{n_k+1})$$
$$= \alpha_{v_k} \cdot \alpha_{n_k} \cdot F\left[\frac{\sqrt{m(x_{n_k}, x_{v_k+1}) \cdot m(x_{v_k}, x_{n_k+1})}}{\psi(m(x_{n_k}, x_{v_k+1}), m(x_{v_k}, x_{n_k+1}))}\right]$$

Since $\lim_{n\to\infty} m(x_{n+1}, x_n) = 1$, $\lim_{n\to\infty} m(x_{n_k}, x_{v_k+1}) = \epsilon$, $\lim_{n\to\infty} m(x_{v_k}, x_{n_k+1}) = \epsilon$, then by the continuity of ψ and the properties of F, if we take limits in the above inequality, as $k \to \infty$, we deduce that

$$\epsilon \leq F\left[\frac{\sqrt{\epsilon^2}}{\psi(\epsilon, \epsilon)}\right] \leq \epsilon$$

From the above we deduce that $\psi(\epsilon, \epsilon) = 1$, and it follows from the properties of ψ, that $\epsilon = 1$, which is a contradiction. Thus, $\{x_n\}$ is a multiplicative Cauchy sequence. Since A is a closed subset of of the complete multiplicative metric space X, there exists $x \in A$ such that $\lim_{n\to\infty} x_n = x$. It follows upon taking limits in $m(x_{n+1}, Tx_n) = m(A, B)$ and using the continuity of T that $m(x, Tx) = m(A, B)$

Corollary D.2 1

Let X be a nonempty set such that $(X, \preceq)$ is a partially ordered set and let (X, m) be a complete multiplicative metric space. Let A and B be nonempty closed subsets of X such that A_0 and B_0 are nonempty. Let $T : A \mapsto B$ satisfy the following conditions

(a) T is continuous, increasing such that $T(A_0) \subseteq B_0$ and $x \preceq y$, $m(u, Tx) = m(A, B)$, $m(v, Ty) = m(A, B)$ implies $m(u, v) \leq [m(x, v) \cdot m(y, u)]^\alpha$ where $\alpha \in (0, \frac{1}{2})$

(b) there exists an element x_0 and x_1 in A_0 such that $x_0 \preceq x_1$ and $m(x_1, Tx_0) = m(A, B)$

Then there exists a point $x \in A$ such that $m(x, Tx) = m(A, B)$. Moreover, for any fixed $x_0 \in A_0$, the sequence $\{x_n\}$ defined by $m(x_{n+1}, Tx_n) = m(A, B)$ converges to the point x

Proof of Corollary D.2 1

Let $\alpha \in (0, \frac{1}{2})$, and in the previous theorem take the multiplicative c-class function as $F(x, y) := F\left(\frac{x}{y}\right) = \frac{x}{y}$ and let ψ be given by $\psi(a, b) = (a \cdot b)^{\frac{1}{2} - \alpha}$, then it follows that $\psi(a, b) = 1$ iff $a = b = 1$. Moreover in this case, the contractive condition of the previous theorem reduces to the one contained in the above Corollary

For a self-mapping condition (b) of the previous theorem implies $x_0 \preceq Tx_0$, thus we get the following which is a multiplicative generalization of a result due to Harjani et al [Harjani, J, López, B, Sadarangani, K: Fixed point theorems for weakly C-contractive mappings in ordered metric spaces. Comput. Math. Appl. 61, 790-796 (2011)]

CHAPTER 4. BEST PROXIMITY POINT THEOREMS FOR IMPLICIT GENERALIZED PROXIMAL C-MULTIPLICATIVE CONTRACTION WITH PARTIAL ORDERS

Corollary D.3 1

Let X be a nonempty set such that $(X, \preceq)$ is a partially ordered set and let (X, m) be a complete multiplicative metric space. Let $T : X \mapsto X$ be a continuous and nondecreasing mapping such that, for all $x, y \in X$,

$$m(Tx, Ty) \leq F\left[\frac{\sqrt{m(x, Ty) \cdot m(y, Tx)}}{\psi(m(x, Ty), m(y, Tx))}\right]$$

for $x \preceq y$, where $\psi : [1, \infty)^2 \mapsto [1, \infty)$ is a continuous and nondecreasing function such that $\psi(x, y) = 1$ iff $x = y = 1$, and $F(x, y) := F\left(\frac{x}{y}\right)$ is a multiplicative C-class function [Clement Ampadu and Arslan Hojat Ansari, FIXED POINT THEOREMS IN COMPLETE MULTIPLICATIVE METRIC SPACES WITH APPLICATION TO MULTIPLICATIVE ANALOGUE OF C-CLASS FUNCTIONS, JP Journal of Fixed Point Theory and Applications, August 2016, Volume 11, Issue 2, Pages 113-124]. If there exists $x_0 \in X$ with $x_0 \preceq Tx_0$, then T has a fixed point

If we drop the continuity condition on T in the previous theorem, then we obtain the following

Theorem D.4 1

Let X be a nonempty set such that $(X, \preceq)$ is a partially ordered set and let (X, m) be a complete multiplicative metric space. Let A and B be nonempty closed subsets of X such that A_0 and B_0 are nonempty. Let $T : A \mapsto B$ satisfy the following conditions

(a) T is a proximally order-preserving and implicit generalized proximal C- multiplicative contraction such that $T(A_0) \subseteq B_0$

(b) there exists an element x_0 and x_1 in A_0 such that $x_0 \preceq x_1$ and $m(x_1, Tx_0) = m(A, B)$

(c) if $\{x_n\}$ is an increasing sequence in A converging to x, then $x_n \preceq x$ for all $n \in \mathbb{N}$

Then there exists a point $x \in A$ such that $m(x, Tx) = m(A, B)$.

Proof of Theorem D.4 1

As in the proof of the previous theorem, we have $m(x_{n+1}, Tx_n) = m(A, B)$ for all $n \geq 0$. Moreover, $\{x_n\}$ is a multiplicative Cauchy sequence and converges to some point $x \in A$. Now observe for each $n \in \mathbb{N}$ we have

$$m(A, B) = m(x_{n+1}, Tx_n)$$
$$\leq m(x_{n+1}, x) \cdot m(x, Tx_n)$$
$$\leq m(x, x_{n+1})^2 \cdot m(x_{n+1}, Tx_n)$$
$$\leq m(x, x_{n+1})^2 \cdot m(A, B)$$

Taking limits in the above inequality, we deduce that $\lim_{n \to \infty} m(x, Tx_n) = m(A, B)$ and hence $x \in A_0$. Since $T(A_0) \subseteq B_0$, there exists $v \in A$ such that $m(v, Tx) = m(A, B)$. Now we prove that $x = v$. By condition (c) of the theorem we have $x_n \preceq x$ for all $n \in \mathbb{N}$. Now since $m(x_{n+1}, Tx_n) = m(A, B)$, $m(v, Tx) = m(A, B)$, and T is an implicit generalized proximal C-multiplicative contraction, we deduce that

$$m(x_{n+1}, v) \leq F\left[\frac{\sqrt{m(x_n, v) \cdot m(x, x_{n+1})}}{\psi(m(x_n, v), m(x, x_{n+1}))}\right]$$

Now taking limits in the above, we deduce that

$$m(x, v) \leq F\left[\frac{\sqrt{m(x, v)}}{\psi(m(x, v), 1)}\right]$$

from which it follows that $m(x, v) = 1$, that is $x = v$. If we replace v by x in $m(v, Tx) = m(A, B)$, we obtain $m(x, Tx) = m(A, B)$

Let $\alpha \in \left(0, \frac{1}{2}\right)$, and in the previous theorem take the multiplicative c-class function as $F(x, y) := F\left(\frac{x}{y}\right) = \frac{x}{y}$ and let ψ be given by $\psi(a, b) = (a \cdot b)^{\frac{1}{2} - \alpha}$, then it follows that $\psi(a, b) = 1$ iff $a = b = 1$. Moreover in this case, the contractive condition of the previous theorem reduces to the one contained below

Corollary D.5 1

Let X be a nonempty set such that $(X, \preceq)$ is a partially ordered set and let (X, m) be a complete multiplicative metric space. Let A and B be nonempty closed subsets of X such that A_0 and B_0 are nonempty. Let $T : A \mapsto B$ satisfy the following conditions

(a) T is an increasing mapping such that $T(A_0) \subseteq B_0$ and $x \preceq y$, $m(u, Tx) = m(A, B)$, $m(v, Ty) = m(A, B)$ implies $m(u, v) \leq [m(x, v) \cdot m(y, u)]^\alpha$ where $\alpha \in \left(0, \frac{1}{2}\right)$

(b) there exists an element x_0 and x_1 in A_0 such that $x_0 \preceq x_1$ and $m(x_1, Tx_0) = m(A, B)$

(c) if $\{x_n\}$ is an increasing sequence in A converging to a point $x \in X$, then $x_n \preceq x$ for all $n \in \mathbb{N}$

Then there exists a point $x \in A$ such that $m(x, Tx) = m(A, B)$.

For a self-mapping condition (b) of the previous theorem implies $x_0 \preceq Tx_0$, thus we get the following which is a multiplicative generalization of a result due to Harjani et al [Harjani, J, López, B, Sadarangani, K: Fixed point theorems for weakly C-contractive mappings in ordered metric spaces. Comput. Math. Appl. 61, 790-796 (2011)]

CHAPTER 4. BEST PROXIMITY POINT THEOREMS FOR IMPLICIT GENERALIZED PROXIMAL C-MULTIPLICATIVE CONTRACTION WITH PARTIAL ORDERS

> **Corollary D.6 1**
>
> Let X be a nonempty set such that $(X, \preceq)$ is a partially ordered set and let (X, m) be a complete multiplicative metric space. Assume that if $\{x_n\} \subseteq X$ is a nondecreasing sequence such that $x_n \to x$ in X, then $x_n \preceq x$ for all $n \in \mathbb{N}$. Let $T : X \mapsto X$ be nondecreasing mapping such that
>
> $$m(Tx, Ty) \leq F\left[\frac{\sqrt{m(x, Ty) \cdot m(y, Tx)}}{\psi(m(x, Ty), m(y, Tx))}\right]$$
>
> for $x \preceq y$, where $\psi : [1, \infty)^2 \mapsto [1, \infty)$ is a continuous and nondecreasing function such that $\psi(x, y) = 1$ iff $x = y = 1$, and $F(x, y) := F\left(\frac{x}{y}\right)$ is a multiplicative C-class function [Clement Ampadu and Arslan Hojat Ansari, FIXED POINT THEOREMS IN COMPLETE MULTIPLICATIVE METRIC SPACES WITH APPLICATION TO MULTIPLICATIVE ANALOGUE OF C-CLASS FUNCTIONS, JP Journal of Fixed Point Theory and Applications, August 2016, Volume 11, Issue 2, Pages 113-124]. If there exists $x_0 \in X$ with $x_0 \preceq Tx_0$, then T has a fixed point

For uniqueness of the best proximity point in the previous two theorems of this chapter we need the following [Nieto, JJ, Rodríguez-López, R: Contractive mapping theorems in partially ordered sets and applications to ordinary differential equations. Order 22, 223-239 (2005)]: For $x, y \in X$, there exists $z \in X$ which is comparable to x and y. In particular under this condition, we can guarantee uniqueness of the best proximity point in Theorem D.1 as follows

> **Theorem D.7 1**
>
> Let X be a nonempty set such that $(X, \preceq)$ is a partially ordered set and let (X, m) be a complete multiplicative metric space. Let A and B be nonempty closed subsets of X and let A_0 and B_0 be nonempty such that A_0 satisfies the condition:
>
> for $x, y \in X$, there exists $z \in X$ which is comparable to x and y
>
> Let $T : A \mapsto B$ satisfy the following conditions
>
> (a) T is a continuous, proximally order preserving and implicit generalized proximal C-multiplicative contraction such that $T(A_0) \subseteq B_0$
>
> (b) there exists an element x_0 and x_1 in A_0 such that $x_0 \preceq x_1$ and $m(x_1, Tx_0) = m(A, B)$
>
> Then there exists a unique point $x \in A$ such that $m(x, Tx) = m(A, B)$

CHAPTER 4. BEST PROXIMITY POINT THEOREMS FOR IMPLICIT GENERALIZED PROXIMAL C-MULTIPLICATIVE CONTRACTION WITH PARTIAL ORDERS

Proof of Theorem D.7 1

It is sufficient to show the uniqueness of the point $x \in A$ such that $m(x, Tx) = m(A, B)$. Suppose there exists x and x^* in A which are best proximity points, that is, $m(x, Tx) = m(A, B)$ and $m(x^*, Tx^*) = m(A, B)$

Case I: x is comparable to x^*, that is, $x \preceq x^*$ (or $x^* \preceq x$)

Since T is an implicit generalized proximal C-multiplicative contraction, we deduce the following

$$m(x, x^*) \leq F\left[\frac{\sqrt{m(x, x^*) \cdot m(x^*, x)}}{\psi(m(x, x^*), m(x^*, x))}\right] \leq m(x, x^*)$$

from which it follows that $\psi(m(x, x^*), m(x^*, x)) = 1$ and thus $m(x, x^*) = 1$, that is, $x = x^*$

Case II: x is not comparable to x^*

Since A_0 satisfies the condition, "for $x, y \in X$, there exists $z \in X$ which is comparable to x and y," then there exists $z \in A_0$ such that z is comparable to x and x^*, that is, $x \preceq z$ (or $z \preceq x$) and $x^* \preceq z$ (or $z \preceq x^*$). Suppose that $x \preceq z$ and $x^* \preceq z$. Since $T(A_0) \subseteq B_0$, there exists a point $v_0 \in A_0$ such that $m(v_0, Tz) = m(A, B)$. By the proximally order-preserving property of T, we get $x \preceq v_0$ and $x^* \preceq v_0$. Since $T(A_0) \subseteq B_0$, there exists a point $v_1 \in A_0$ such that $m(v_1, Tv_0) = m(A, B)$. Again by the proximally order-preserving property of T, we get $x \preceq v_1$ and $x^* \preceq v_1$. Proceeding in a similar fashion, we can find $v_n \in A_0$ with $v_{n+1} \in A_0$ such that $m(v_{n+1}, Tv_n) = m(A, B)$. Hence $x \preceq v_n$ and $x^* \preceq v_n$ for all $n \in \mathbb{N}$. Since T is an implicit generalized proximal C-multiplicative contraction, we deduce that

$$m(v_{n+1}, x) \leq F\left[\frac{\sqrt{m(v_n, x) \cdot m(x, v_{n+1})}}{\psi(m(v_n, x), m(x, v_{n+1}))}\right]$$

and

$$m(v_{n+1}, x^*) \leq F\left[\frac{\sqrt{m(v_n, x^*) \cdot m(x^*, v_{n+1})}}{\psi(m(v_n, x^*), m(x^*, v_{n+1}))}\right]$$

If we take limits as $n \to \infty$ in the two inequalities immediately above we deduce that $\lim_{n \to \infty} v_n = x$ and $\lim_{n \to \infty} v_n = x^*$. Thus, by the uniqueness of the limit, we conclude that $x = x^*$. Other cases can be proved similarly.

For uniqueness of the best proximity point in Theorem D.4, we have the following

Theorem D.8 1

Let X be a nonempty set such that $(X, \preceq)$ is a partially ordered set and let (X, m) be a complete multiplicative metric space. Let A and B be nonempty closed subsets of X and let A_0 and B_0 be nonempty such that A_0 satisfies the condition:

for $x, y \in X$, there exists $z \in X$ which is comparable to x and y

Let $T : A \mapsto B$ satisfy the following conditions

(a) T is a proximally order preserving and implicit generalized proximal C-multiplicative contraction such that $T(A_0) \subseteq B_0$

(b) there exists an element x_0 and x_1 in A_0 such that $x_0 \preceq x_1$ and $m(x_1, Tx_0) = m(A, B)$

(c) if $\{x_n\}$ is an increasing sequence in A converging to x, then $x_n \preceq x$ for all $n \in \mathbb{N}$

Then there exists a unique point $x \in A$ such that $m(x, Tx) = m(A, B)$

4.4 Exercises

Exercise D.1 1

Taking inspiration from [Chirasak Mongkolkeha, Yeol Je Cho and Poom Kumam, Best proximity points for generalized proximal C-contraction mappings in metric spaces with partial orders, Journal of Inequalities and Applications 2013, 2013:94] give an example to support Theorem D.1

CHAPTER 4. BEST PROXIMITY POINT THEOREMS FOR IMPLICIT GENERALIZED PROXIMAL C-MULTIPLICATIVE CONTRACTION WITH PARTIAL ORDERS

> **Exercise D.2 1**
>
> A mapping $T : A \mapsto B$ will be called an implicit generalized proximal Kannan-multiplicative contraction, if for all $u, v, x, y \in A$, it satisfies
>
> $$x \preceq y, m(u, Tx) = m(A, B), m(v, Ty) = m(A, B)$$
>
> implies
>
> $$m(u, v) \leq F\left[\frac{\sqrt{m(x, u) \cdot m(y, v)}}{\psi(m(x, u), m(y, v))}\right]$$
>
> where $\psi : [1, \infty)^2 \mapsto [1, \infty)$ is a continuous and nondecreasing function such that $\psi(x, y) = 1$ iff $x = y = 1$, and $F(x, y) := F\left(\frac{x}{y}\right)$ is a multiplicative C-class function [Clement Ampadu and Arslan Hojat Ansari, FIXED POINT THEOREMS IN COMPLETE MULTIPLICATIVE METRIC SPACES WITH APPLICATION TO MULTIPLICATIVE ANALOGUE OF C-CLASS FUNCTIONS, JP Journal of Fixed Point Theory and Applications, August 2016, Volume 11, Issue 2, Pages 113-124].
>
> (i) Prove the following: Let X be a nonempty set such that $(X, \preceq)$ is a partially ordered set and let (X, m) be a complete multiplicative metric space. Let A and B be nonempty closed subsets of X such that A_0 and B_0 are nonempty. Let $T : A \mapsto B$ satisfy the following conditions
>
>   (a) T is a continuous, proximally order-preserving and implicit generalized proximal Kannan-multiplicative contraction such that $T(A_0) \subseteq B_0$
>
>   (b) there exists an element x_0 and x_1 in A_0 such that $x_0 \preceq x_1$ and $m(x_1, Tx_0) = m(A, B)$
>
>   Then there exists a point $x \in A$ such that $m(x, Tx) = m(A, B)$. Moreover, for any fixed $x_0 \in A_0$, the sequence $\{x_n\}$ defined by $m(x_{n+1}, Tx_n) = m(A, B)$ converges to the point x
>
> (ii) Prove the following: Let X be a nonempty set such that $(X, \preceq)$ is a partially ordered set and let (X, m) be a complete multiplicative metric space. Let A and B be nonempty closed subsets of X and let A_0 and B_0 be nonempty such that A_0 satisfies the condition:
>
>   for $x, y \in X$, there exists $z \in X$ which is comparable to x and y
>
>   Let $T : A \mapsto B$ satisfy the following conditions
>
>   (a) T is a continuous, proximally order preserving and implicit generalized proximal Kannan-multiplicative contraction such that $T(A_0) \subseteq B_0$
>
>   (b) there exists an element x_0 and x_1 in A_0 such that $x_0 \preceq x_1$ and $m(x_1, Tx_0) = m(A, B)$
>
>   Then there exists a unique point $x \in A$ such that $m(x, Tx) = m(A, B)$
>
> (iii) Let $\alpha \in \left(0, \frac{1}{2}\right)$, and in (i) take the multiplicative c-class function as $F(x, y) := F\left(\frac{x}{y}\right) = \frac{x}{y}$ and let ψ be given by $\psi(a, b) = (a \cdot b)^{\frac{1}{2} - \alpha}$, then it follows that $\psi(a, b) = 1$ iff $a = b = 1$. State the Corollary arising from (i) in this case

Exercise D.3 1

A mapping $T : A \mapsto B$ will be called an implicit generalized proximal Reich type-multiplicative contraction, if for all $u, v, x, y \in A$, it satisfies

$$x \preceq y, m(u, Tx) = m(A, B), m(v, Ty) = m(A, B)$$

implies

$$m(u, v) \leq F\left[\frac{\sqrt[3]{m(x, u) \cdot m(y, v) \cdot m(x, y)}}{\psi(m(x, u), m(y, v), m(x, y))}\right]$$

where $\psi : [1, \infty)^3 \mapsto [1, \infty)$ is a continuous and nondecreasing function such that $\psi(x, y, z) = 1$ iff $x = y = z = 1$, and $F(x, y) := F(\frac{x}{y})$ is a multiplicative C-class function [Clement Ampadu and Arslan Hojat Ansari, FIXED POINT THEOREMS IN COMPLETE MULTIPLICATIVE METRIC SPACES WITH APPLICATION TO MULTIPLICATIVE ANALOGUE OF C-CLASS FUNCTIONS, JP Journal of Fixed Point Theory and Applications, August 2016, Volume 11, Issue 2, Pages 113-124].

(i) Prove the following: Let X be a nonempty set such that $(X, \preceq)$ is a partially ordered set and let (X, m) be a complete multiplicative metric space. Let A and B be nonempty closed subsets of X such that A_0 and B_0 are nonempty. Let $T : A \mapsto B$ satisfy the following conditions

(a) T is a continuous, proximally order-preserving and implicit generalized proximal Reich type-multiplicative contraction such that $T(A_0) \subseteq B_0$

(b) there exists an element x_0 and x_1 in A_0 such that $x_0 \preceq x_1$ and $m(x_1, Tx_0) = m(A, B)$

Then there exists a point $x \in A$ such that $m(x, Tx) = m(A, B)$. Moreover, for any fixed $x_0 \in A_0$, the sequence $\{x_n\}$ defined by $m(x_{n+1}, Tx_n) = m(A, B)$ converges to the point x

(ii) Prove the following: Let X be a nonempty set such that $(X, \preceq)$ is a partially ordered set and let (X, m) be a complete multiplicative metric space. Let A and B be nonempty closed subsets of X and let A_0 and B_0 be nonempty such that A_0 satisfies the condition:

for $x, y \in X$, there exists $z \in X$ which is comparable to x and y

Let $T : A \mapsto B$ satisfy the following conditions

(a) T is a continuous, proximally order preserving and implicit generalized proximal Reich type-multiplicative contraction such that $T(A_0) \subseteq B_0$

(b) there exists an element x_0 and x_1 in A_0 such that $x_0 \preceq x_1$ and $m(x_1, Tx_0) = m(A, B)$

Then there exists a unique point $x \in A$ such that $m(x, Tx) = m(A, B)$

(iii) Let $\alpha \in \left(0, \frac{1}{3}\right)$, and in (i) take the multiplicative c-class function as $F(x, y) := F(\frac{x}{y}) = \frac{x}{y}$ and let ψ be given by $\psi(a, b, c) = (a \cdot b \cdot c)^{\frac{1}{3} - \alpha}$, then it follows that $\psi(a, b, c) = 1$ iff $a = b = c = 1$. State the Corollary arising from (i) in this case

CHAPTER 4. BEST PROXIMITY POINT THEOREMS FOR IMPLICIT GENERALIZED PROXIMAL C-MULTIPLICATIVE CONTRACTION WITH PARTIAL ORDERS

> **Exercise D.4 1**
>
> A mapping $T : A \mapsto B$ will be called an implicit generalized proximal Hardy and Rogers type-multiplicative contraction, if for all $u, v, x, y \in A$, it satisfies
>
> $$x \preceq y, m(u, Tx) = m(A, B), m(v, Ty) = m(A, B)$$
>
> implies
>
> $$m(u,v) \leq F\left[\frac{\sqrt[5]{m(x,u)\cdot m(y,v)\cdot m(x,y)\cdot m(x,v)\cdot m(y,u)}}{\psi(m(x,u), m(y,v), m(x,y), m(x,v), m(y,u))}\right]$$
>
> where $\psi : [1, \infty)^5 \mapsto [1, \infty)$ is a continuous and nondecreasing function such that $\psi(x, y, z, t, r) = 1$ iff $x = y = z = t = r = 1$, and $F(x, y) := F\left(\frac{x}{y}\right)$ is a multiplicative C-class function [Clement Ampadu and Arslan Hojat Ansari, FIXED POINT THEOREMS IN COMPLETE MULTIPLICATIVE METRIC SPACES WITH APPLICATION TO MULTIPLICATIVE ANALOGUE OF C-CLASS FUNCTIONS, JP Journal of Fixed Point Theory and Applications, August 2016, Volume 11, Issue 2, Pages 113-124].
>
> (i) Prove the following: Let X be a nonempty set such that $(X, \preceq)$ is a partially ordered set and let (X, m) be a complete multiplicative metric space. Let A and B be nonempty closed subsets of X such that A_0 and B_0 are nonempty. Let $T : A \mapsto B$ satisfy the following conditions
>
>   (a) T is a proximally order-preserving and implicit generalized proximal Hardy and Rogers type-multiplicative contraction such that $T(A_0) \subseteq B_0$
>
>   (b) there exists an element x_0 and x_1 in A_0 such that $x_0 \preceq x_1$ and $m(x_1, Tx_0) = m(A, B)$
>
>   (c) if $\{x_n\}$ is an increasing sequence in A converging to x, then $x_n \preceq x$ for all $n \in \mathbb{N}$
>
> Then there exists a point $x \in A$ such that $m(x, Tx) = m(A, B)$.
>
> (ii) Prove the following: Let X be a nonempty set such that $(X, \preceq)$ is a partially ordered set and let (X, m) be a complete multiplicative metric space. Let A and B be nonempty closed subsets of X and let A_0 and B_0 be nonempty such that A_0 satisfies the condition:
>
> for $x, y \in X$, there exists $z \in X$ which is comparable to x and y
>
> Let $T : A \mapsto B$ satisfy the following conditions
>
>   (a) T is a proximally order preserving and implicit generalized proximal Hardy and Rogers type-multiplicative contraction such that $T(A_0) \subseteq B_0$
>
>   (b) there exists an element x_0 and x_1 in A_0 such that $x_0 \preceq x_1$ and $m(x_1, Tx_0) = m(A, B)$
>
>   (c) if $\{x_n\}$ is an increasing sequence in A converging to x, then $x_n \preceq x$ for all $n \in \mathbb{N}$
>
> Then there exists a unique point $x \in A$ such that $m(x, Tx) = m(A, B)$
>
> (iii) Let $\alpha \in \left(0, \frac{1}{5}\right)$, and in (i) take the multiplicative c-class function as $F(x,y) := F\left(\frac{x}{y}\right) = \frac{x}{y}$ and let ψ be given by $\psi(a, b, c, d, e) = (a \cdot b \cdot c \cdot d \cdot e)^{\frac{1}{5} - \alpha}$, then it follows that $\psi(a, b, c, d, e) = 1$ iff $a = b = c = d = e = 1$. State the Corollary arising from (i) in this case

4.5 References

(1) Choudhury, BS: Unique fixed point theorem for weak-contractive mappings. Kathmandu Univ. J. Sci. Eng. Technol. 5, 6-13 (2009)

(2) Chatterjea, SK: Fixed point theorems. C. R. Acad. Bulgare Sci. 25, 727-730 (1972)

(3) Clement Ampadu and Arslan Hojat Ansari, FIXED POINT THEOREMS IN COMPLETE MULTIPLICATIVE METRIC SPACES WITH APPLICATION TO MULTIPLICATIVE ANALOGUE OF C-CLASS FUNCTIONS, JP Journal of Fixed Point Theory and Applications, August 2016, Volume 11, Issue 2, Pages 113-124

(4) Sadiq Basha, S: Best proximity point theorems on partially ordered sets. Optim. Lett. (2012). doi:10.1007/s11590-012-0489-1

(5) Harjani, J, López, B, Sadarangani, K: Fixed point theorems for weakly C-contractive mappings in ordered metric spaces. Comput. Math. Appl. 61, 790-796 (2011)

(6) Nieto, JJ, Rodríguez-López, R: Contractive mapping theorems in partially ordered sets and applications to ordinary differential equations. Order 22, 223-239 (2005)

(7) Chirasak Mongkolkeha, Yeol Je Cho and Poom Kumam, Best proximity points for generalized proximal C-contraction mappings in metric spaces with partial orders, Journal of Inequalities and Applications 2013, 2013:94